Wise Animals

Dr Tom Chatfield is a British writer, broadcaster and tech philosopher. His books exploring technology and digital culture – most recently *Wise Animals* – have been published in over thirty languages. He has spoken about AI, tech ethics and the future of authorship at venues ranging from the UK and European Parliaments to Google, Meta, the US National Academy of Sciences and TED Global. He lives in Kent.

Wise Animals

How Technology Has Made Us What We Are

TOM CHATFIELD

PICADOR

First published 2024 by Picador

This edition first published 2025 by Picador
an imprint of Pan Macmillan
The Smithson, 6 Briset Street, London ECIM 5NR
EU representative: Macmillan Publishers Ireland Ltd, 1st Floor,
The Liffey Trust Centre, 117–126 Sheriff Street Upper,
Dublin 1, DOI YC43
Associated companies throughout the world
www.panmacmillan.com

ISBN 978-1-5290-7976-0

Copyright © T&C Chatfield Limited 2024

The right of Tom Chatfield to be identified as the
author of this work has been asserted by him in accordance
with the Copyright, Designs and Patents Act 1988.

'September 1, 1939' is copyright © 1940 W. H. Auden, renewed by the
Estate of W. H. Auden. Used by permission of Curtis Brown, Ltd.

All rights reserved. No part of this publication may be reproduced,
stored in a retrieval system, or transmitted, in any form, or by any means
(electronic, mechanical, photocopying, recording or otherwise)
without the prior written permission of the publisher.

Pan Macmillan does not have any control over, or any responsibility for,
any author or third-party websites referred to in or on this book.

1 3 5 7 9 8 6 4 2

A CIP catalogue record for this book is available from the British Library.

Typeset in Garamond Pro by Jouve (UK), Milton Keynes
Printed and bound by CPI Group (UK) Ltd, Croydon, CR0 4YY

This book is sold subject to the condition that it shall not, by way of
trade or otherwise, be lent, hired out, or otherwise circulated without
the publisher's prior consent in any form of binding or cover other than
that in which it is published and without a similar condition including
this condition being imposed on the subsequent purchaser.

Visit **www.picador.com** to read more about all our books
and to buy them. You will also find features, author interviews and
news of any author events, and you can sign up for e-newsletters
so that you're always first to hear about our new releases.

For Mum, Cat, Toby and Clio. Always.

Contents

Introduction ... 1

PART ONE: Origins

1. Technology in deep time: *The delusion of inevitability* ... 9
2. Stone and fire: *The delusion of mastery* ... 25
3. Love and learning: *The delusion of brutality* ... 43
4. Extended minds: *The delusion of 'it' and 'us'* ... 59
5. Consciousness as controlled hallucination: *The delusion of literal-mindedness* ... 81
6. How technologies invent themselves: *The delusion of comprehension* ... 97

PART TWO: Destinations

7. Values and assumptions: *The delusion of neutrality* ... 121
8. Myths and wish-fulfilment: *The delusion of magical thinking* ... 139
9. Trickery and intellect: *The anthropomorphic delusion* ... 161
10. Superintelligence and doubt: *The delusion of machine perfection* ... 193

CONTENTS

11. Towards a new ethics of technology: *The delusion of divine data*   215
12. Death and life: *The delusion of perpetual progress*   239

And Finally   255

Acknowledgements   259
Thanks   261
Select Bibliography   263
Notes   269
Index   309

The machine itself makes no demands and holds out no promises: it is the human spirit that makes demands and keeps promises. In order to reconquer the machine and subdue it to human processes, one must first understand it and assimilate it.

                      Lewis Mumford, *Technics & Civilisation*

I can only answer the question 'What am I to do?' if I can answer the prior question 'Of what story or stories do I find myself a part?'

                      Alasdair MacIntyre, *After Virtue*

But they are useless. They can only give you answers.

                      Pablo Picasso on computers, *The Paris Review*

Introduction

In his 1999 essay 'How to Stop Worrying and Learn to Love the Internet', the British author Douglas Adams had this to say about technology.

> 'Technology', as the computer scientist Bran Ferren memorably defined it, is 'stuff that doesn't work yet.' We no longer think of chairs as technology, we just think of them as chairs. But there was a time when we hadn't worked out how many legs chairs should have, how tall they should be, and they would often 'crash' when we tried to use them. Before long, computers will be as trivial and plentiful as chairs . . . and we will cease to be aware of the things.[1]

Adams was both being very funny and making a serious point. The world we're born into is underpinned by millennia of innovation and ingenuity, most of which long ago sank beneath the surface of human attention. 'Technology' is the shiny, strange, new stuff; the gadgets we can't wait to get our hands on as teenagers, only to fret over their mind-warping powers once we become parents. So should we all take a deep breath, re-read his essay's title, then stop worrying and accept that this is just how things are?

Despite being a lifelong fan of Adams's writing, my answer is 'no'.

In fact, it's important for us to head in the opposite direction. It's not that the world needs more hand-wringing over smartphones, or the kind of nostalgia that forgets how much suffering our ancestors endured (unreliable seating wasn't the half of it). Rather, we need the right kind of worries. And this means turning our gaze away from gadgets towards the values and assumptions baked into them – not to mention the structures, incentives and understandings surrounding their creation.

Is it a good thing that my children enjoy near-instantaneous access to more knowledge than the entirety of humanity possessed half a century ago? Absolutely. But the fact that a few titanic corporations mediate most of their encounters with this knowledge is less desirable; while the fact that conspiracists, cranks, hate-mongers and professional narcissists are given prominent platforms by some of these same corporations is downright depressing. And that's before you dip your toes into the murky waters of algorithmic inscrutability, bias and surveillance – let alone the future of AI or the ongoing environmental consequences of industrialization.

As Adams notes, most of us treat the technologies that already existed when we were born as normal; those invented between then and our thirtieth birthday as exciting; and anything invented after the age of thirty as 'against the natural order of things and the beginning of the end of civilisation as we know it until it's been around for about ten years when it gradually turns out to be alright really.' As a species, we are ceaselessly in the process of reinventing ourselves, and this adaptability is central to our thriving. But it also makes us vulnerable, not least to the assumption that the cultural and technological order we're born into is both natural and inevitable – while only 'the stuff that doesn't work yet' is worth debating.

If you truly love technology, the opposite is true. To love something is to be obliged to worry about its history, purposes and imperfections; to want it to be better; to say *no* as well as *yes* to its offerings.[2] Caring about technology means paying close attention to many of the things it encourages us to forget: that the world didn't have to be this way; that it isn't going to stay this way for long; and that what happens next is, to a discomforting degree, up to us.

'We shape our tools, and thereafter our tools shape us.' The media theorist Marshall McLuhan didn't actually say this (it was written by his friend Father John Culkin in a 1967 article discussing McLuhan's work) but it captures one of McLuhan's most enduring insights.[3] Technology exists in a constant dance with its creators, each influencing the other, neither able to go it alone.

Faced by the power, complexity and momentum of this inheritance, it's easy to despair: to submit to the sense that, sooner or later, the systems we've created will inexorably save or condemn us.[4] Yet the determinism underpinning this view is flawed at its foundations: blind to the entwining of our own and our creations' evolution; deaf to the lessons that history, biology and art can teach. We are – for better and for worse – both free and obligated to negotiate the terms of our existence.

How should we set about making our worries useful? First, we need to see that the deep histories of our own and our technologies' evolutions are inextricably entwined, and that we can't hope to understand one in the absence of the other. Second, we need to resist the denigration of human agency in the face of our creations' scale and significance: to acknowledge the depths of our interdependence while refusing to accept that there's anything inevitable about what lies ahead.

In particular, it's vital for us to overcome certain *delusions* when it comes to our conceptions of technology: false beliefs that stand between us and a rich, reality-based engagement with the twenty-first century's challenges.

Technology is everywhere today. It touches everything we do and believe ourselves to be. It shapes our deepest anxieties and hopes, our politics and our most intimate relationships. Yet many of the most influential stories through which we seek to understand it are told from the wrong angle: as if its nature can be debated without reference to our own; as if people and machines are locked in an existential struggle over everything from work and leisure to love and art; as if technology's progress must necessarily define humanity's.

None of this is true, and much of it is harmful. Technology isn't just something we make or do to the world; that we pick up or put down. We cannot separate ourselves from it, because it has been with us since before the beginning, evolving alongside us, shaping our biology and our ecology. Contrary to many wishful critiques, there is no such thing as human nature or existence in the absence of technology; and this was already true when *Homo sapiens* first walked the African continent over three hundred thousand years ago. Nor is there any such thing as a neutral tool, untouched by human designs and desires.

Our species is wise and foolish in ways inconceivable to any other creature – and technology is implicated in every one of them. Thanks to the labour of countless generations, we can collectively describe, explain and remake our world in remarkable ways; can shield ourselves against brute necessity while dreaming new selves. We can apprehend our universe's vastest and most microscopic scales while mourning the transience of each individual life. Yet we can also be prodigiously self-defeating, solipsistic and destructive; deniers and deceivers as much as creators. Confronting the future hopefully means finding a way to acknowledge all of these things: our vertiginous ambition, scope and vulnerability.

At least at the time of writing, machine superintelligences hadn't entered the picture. But the faith that our fate rests in the hands of a technocratic elite is alive and well, alongside the delusional hopes and discontents it breeds: the abnegation of collective responsibility; the miscasting of certain individuals and corporations as avatars of destiny. Against this, we need a conception of the human-made world that acknowledges its continuities with this planet's other systems; that embraces the virtues of compassion, curiosity and humility; and that promises us neither certainty nor mastery, but rather the collective struggle to become less deceived.[5]

PART ONE

Origins

CHAPTER 1

Technology in deep time:
The delusion of inevitability

Technology, in the sense I'll be exploring over the course of this book, describes the entirety of the human-made artefacts that extend our grasp of the world: not just computers and cars and planes, but also clocks, clothes, dwellings, weapons, written words and cooking vessels.[1] We can't be sure when the first such artefacts appeared. But we do know that over three million years ago, ancestors so distant they didn't even belong to our genus were creating crude tools: stones that they sharpened, in order, most likely, to help with butchery and breaking open animal bones.[2]

In this, these ancient ancestors weren't so different from several other groups of animals living today. Plenty of creatures can communicate richly, comprehend one another's intentions and put tools to intelligent and creative use: not only our fellow primates but also cetaceans (whales and dolphins), cephalopods (squids and octopuses) and corvids (crows and ravens). Some can even develop and pass on particular local practices. Subpopulations of orcas have been observed developing and passing on social 'fads' such as ramming boats, while New Caledonian crows exhibit a 'culture' of tool usage, creating distinct varieties of simple hooked tools from leaves in order to help them feed on insects.[3]

Humans, however, are unique in having turned such craft into a shared body of knowledge and practices: a system of collective, cumulative discoveries that over mere hundreds of thousands of years has harnessed phenomena like fire to cook food, smelt metal and generate power; gravity into systems of levers, ramps, pulleys, wheels and counterweights; mental processes into art, numeracy, literacy and computation.

This, above all, marks humanity's departure from the rest of life on Earth. Alone among species – at least until the crows have put in a few million years' more effort – humans can improve and combine their creations over time. It is through this process of recursive iteration that tools became technologies; and technology a world-altering force.

In his 2009 book *The Nature of Technology*, the economist W. Brian Arthur argues that it's not only pointless but also actively misleading to do what most history books cannot resist, and treat technology's lineage as a greatest-hits list of influential inventions and inventors: to tell stirring tales of the creation of the compass, the clock, the printing press, the lightbulb, the iPhone. This is not because such inventions weren't important, but because it obscures the fact that all new technologies are at root a *recombination* of older technologies – and that their emergence enacts an evolutionary process resembling the one governing life itself.

Consider the printing press, the inevitable poster-child for anyone wanting to offer a historical-ish perspective on the dissemination of information. The German inventor Johannes Gutenberg was, famously, the first European to develop a system for printing with movable type, in around 1450. Yet he was far from the first person to realize that using individual, movable components to create each character in a line was a good way to speed up printing.

Indeed, Gutenberg's heroic presence in Western cultural history – complete with eponymous Bible – obscures many of the most significant elements of print's global story. As early as the year 800, wooden blocks dipped in ink were being used in China to print entire pages of text onto paper. Between 972 and 983, no fewer than 5,048 volumes of the Buddhist canon known as the *Tripitaka* were printed from such wood blocks in the city of Chengdu, totalling around 130,000 unique pages. Experiments with movable wood-chiselled ideograms followed, along with attempts at producing individual porcelain characters, knowledge that spread alongside Chinese imperial expansion.[4]

By around 1234, the Korean civil minister and scholar Choe Yun-ui – tasked by the ruling Goryeo dynasty with producing multiple copies of the Buddhist text *The Prescribed Ritual Text of the Past and Present* (*Sangjeong Gogeum Yemun*) – had adapted a coin-minting technique to create the first known form of printing with bronze characters, which were held in a wooden frame, inked and pressed against paper. Choe Yun-ui is, arguably, the closest thing movable metal type has to an 'inventor'; but the books he created were not widely distributed.

By the time the aforementioned German craftsman Johannes Gutenberg began seeking investors for his latest venture two centuries later, some knowledge of printing had likely travelled along the great east–west trade routes of the silk road. Gutenberg – who had already been involved in a failed venture to make polished metal mirrors for pilgrims – suffered bankruptcy and repeated lawsuits during the course of his efforts to mechanize the labour-intensive business of producing books. But he also benefited from the relatively small number of letters in German; from his knowledge of metal-smelting as a blacksmith and goldsmith, which helped him perfect the casting of a malleable yet durable alloy of lead, tin and antimony; and from his insight that the kind of wooden presses used for centuries in Germany to make wine could be repurposed for pressing type against paper, itself a technology developed in China 1,500 years previously.[5]

Wooden wine-presses, metal alloys, the Roman alphabet, oil-based ink, paper: every piece of the puzzle assembled by Gutenberg and his collaborators was based in a pre-existing technology whose origin could itself be traced back through previous technologies, in unbroken sequence, to the very first tools. We may admire his ingenuity and tenacity – as we should that of Choe Yun-ui – but the desire to valorize a particular moment and individual tells us more about present priorities and preferences than about how technology develops over time.

In a sense this is self-evident. It is, after all, only possible to build something out of components that exist – and these components must, in turn, have been assembled from other pre-existing components, those from others that came before, and so on. Equally self-evidently, this accumulative combination is not by itself sufficient to explain technology's evolution. Another force is required to drive it, and it echoes the one driving the evolution of life itself.

In the case of living things, evolution is based upon a combination of reproduction, selection pressure and heritable variation. The genetic code of successful organisms is passed on, while less successful ones fall by the wayside. Genetic mutations produce incremental variations in species, some of which may prove favourable; while mechanisms such as sexual reproduction combine the genes of different individuals and potentially produce further advantages. Other mechanisms for the recombination of genes include micro-organisms like bacteria adapting themselves to exist entirely inside other organisms' cells, symbiotically conferring benefits upon their hosts.

In the case of technology, survival and reproduction are similarly entwined, but via different underlying mechanisms. This is because technology's transmission has two distinct requirements: the ongoing existence of a species capable of manufacturing it; and (in the case of all human technologies more complex than sticks and stones) networks of supply and maintenance capable of serving technology's own evolving needs.

Humans' fundamental needs are obvious enough – survival and reproduction, based upon adequate food, water and shelter – but in what sense can technology be said to have *needs*? The answer lies all around us, in the immense interlinked ecology of the human-made world. Our creations require power, fuel, raw materials; globe-spanning networks of information, trade and transportation; the creation and maintenance of accrued layers of components that, precisely because they cannot reproduce or repair themselves, bring with them a list of requirements outstripping anything natural. And beneath all this lies the most fundamental fact of all. Technologies *need to be needed*: to fulfil (or, at least, to become bound up with) some human purpose. Each and every one must be conceived, created and prove worth preserving.

Consider the printing press once again. Wine-presses, smelted metal, paper, ink: the moment was ripe for a new technology to combine these and other elements. And it was ripe partly because sufficient interconnections of manufacture and supply existed to make their combination feasible – and scalable. The paper Gutenberg used to print his Bible was imported from the paper-making centre of Caselle in Piedmont, now a part of northern Italy. Its delivery entailed transfer across the Alps by ox cart, then by barge along the waterways of the Rhine. Caselle's expertise had in turn been learned from southern Italy, which had acquired it from Spain and North Africa, whose Muslim rulers had first brought knowledge of paper-making along the silk road from China.[6]

Power, politics and profits all played their part in creating these possibilities: lines of ambition and desire traced across the Earth. So did chance, contingency and the momentum of unfolding events. Yet no matter how complex the context, no matter how unforeseen its consequences, every one of its elements also depended upon the actions or inactions of a human life.

Agriculture and civilization

In its separateness from yet reliance upon biological life, technology is uniquely powerful but also uniquely needy. Made rather than grown, it is unshackled from the limits of flesh and blood. Its capacities are orders of magnitude greater than anything obliged to balance growth and self-preservation, but these capacities rely upon an ever-expanding network of dependencies. In this sense, technology invents many more needs than it serves – with both its requirements and its potentials growing at an exponential rate compared to our own. This exponentially increasing complexity is perhaps technology's most familiar feature. And one of the most familiar stories of its origins is bound up with the emergence, around twelve thousand years ago in a region known as the Fertile Crescent, of agriculture.

It was here, at the heart of the modern Middle East, that as the last ice age retreated hunter-gathering tribes began the gradual process of domesticating the plants and animals they relied upon for sustenance. Wheat, peas, chickpeas and flax were among the first crops bred selectively from wild stock for better yields; pigs, then sheep, then cattle, were similarly selected across thousands of years for qualities such as tameness, milk yield, body fat and bulk.[7] Gradually, humanity was developing the capacity to reshape not only the physical environment but also the genes and natures of the creatures living within it. Human desires had become decisive evolutionary pressures for other species – which, in effect, became biological tools bent to human purposes.

As the global climate became more hospitable, and as tribes travelled and mingled, a swelling variety of human populations across the globe domesticated local flora and fauna, in the process developing ever-expanding bodies of expertise. Multiple elements of agriculture were developed several times across different locations, reflecting a standard feature of evolution: that sufficiently effective innovations can emerge independently across distinct lineages.[8]

New kinds of complexity went hand in hand with these developments. More permanent settlements and more secure food supplies permitted – and demanded – more elaborate building, craft and trading practices than a nomadic existence. By just over five thousand years ago, the so-called Bronze Age (an epoch named for its defining technology, the smelting of tin and copper into the first alloy to be worked into tools) saw the birth of the first civilizations to count their citizens by the tens of thousands, and to build the prototypes of nation states ruled by central authority: Sumer, Ancient Egypt, the Indus Valley, Ancient China.[9]

As these civilizations emerged, the networks of need and expertise surrounding them expanded still further. Great systems of law, commerce, architecture and fealty arose. The kings of Egypt's Early Dynastic period, for example, unified the societies of the Upper Nile and established a capital at Memphis that commanded trade routes to the peoples of the Levant; that grew and stored grain by the tonne; that bought and bartered thousands of agricultural and artisan products while erecting monuments to its monarchs that would endure for millennia. A single family was little more than a cog in the state's machinery.[10]

This is a skeletal narrative of what we now call civilization's emergence: one in which countless technologies played a vital role. It's a familiar tale in essence, if not in detail, with a gathering momentum to its discoveries – and a foreshadowing of all that 'civilization' was to become. Yet the closer you look, the more complex the narrative of cause and effect becomes; and the more any suggestion of a linear progression from 'primitive' to 'civilized' turns out to misrepresent the sheer variety, strangeness and uncertainty of our species' development.

For example, the claim that agriculture and city-states were prerequisites for large-scale construction projects is impossible to reconcile with the fact that the first appearances of monumental architecture in Eurasia substantially *predate* these things. Among the most famous such sites is a vast complex of enclosures in south-east Turkey at a place now known as Göbekli Tepe. Here, from around 11,500 years ago, limestone pillars – weighing up to eight tonnes, standing up to five metres high and hewn into anthropomorphic 'T' shapes – began to be excavated and transported from nearby quarries, then erected in carefully shaped slots and linked by concentric stone walls. Some of these pillars are carved with representations of animals or abstract patterns. All would have required dozens of people working cooperatively for months to excavate and raise, while the site itself has an overarching geometrical scheme that seems to have been built in three distinct phases across approximately 1,500 years of continuous use.

What was the Göbekli Tepe complex *for*? We can never know the details. But we can be certain that a site this elaborate, beautiful and enduring embodied something other than survival and brute necessity; and that it stands as an extraordinary monument to prehistoric humanity's ingenuity and cultural richness. It also suggests a startling inversion of the claim that social complexity necessarily arose from technological innovation. The first great stones of the site were erected in around 9,500 BCE. Nearby, some of the very first evidence of crop cultivation has been traced to half a millennium later. Hunter-gatherers, it seems, came together to erect a vast monument – and then found new ways of feeding the multitudes it drew.[11] As the late, pioneering German archaeologist Klaus Schmidt has argued, excavations at sites like Göbekli Tepe make it increasingly clear that:

the factor that allowed the formation of large, permanent communities was the facility to use symbolic culture, a kind of pre-literate capacity for producing and 'reading' symbolic material culture, that enabled communities to formulate their shared identities, and their cosmos . . . the general function of the enclosures remains mysterious; but it is clear that the pillar statues in the centre of these enclosures represented very powerful beings. If gods existed in the minds of Early Neolithic people, there is an overwhelming probability that the T-shape is the first known monumental depiction of gods.[12]

If Göbekli Tepe were the only site of its kind, it might seem an enigma, drifting outside the main current of history. But mounting evidence suggests it was of a piece with numerous other grand, collective achievements by pre-agricultural societies. Elaborate stone sculptures, carvings and totems erected on a monumental scale across the Levant and Upper Mesopotamia; a wooden totem almost three metres tall found east of the Urals and dated to 12,000 years ago, carved from a single larch and decorated with faces and limbs;[13] even, from as long as 25,000 years ago, a twelve-metre-wide circular structure built from the skeletons of over sixty woolly mammoths near the banks of the River Don in western Russia:[14] these achievements seem to have been symbolically and aesthetically significant, and imply sophisticated shared traditions. And all of this existed outside of the grand narrative of cities, nations, laws and monarchs often treated as synonymous with culture and mass collaboration.[15]

Indeed, the very notion of a steady transition from hunter-gathering to agriculture is hard to reconcile with the actual evidence of cultural practices that survives: of a rich variety of proto-farming, herding and foraging activities that seem to have been developed for equally varied reasons; of ancestors who moved back and forth between differing social and subsistence structures. Instead of clear thresholds, categories and points of no return, we find blurred lines everywhere we look: microcosms of technological and cultural co-evolution that speak to a plethora of possible futures.[16]

What are we to make of this history: of the possibilities dimly glimpsed in archaeological remains? Among other things, it suggests the futility of telling any story about our species' technological development without also delving into the intricacies of thought, belief and culture that drove it; of beauty, artistry and faith. Any narrative we project onto prehistoric times will necessarily be as much about us as it is about those who lived then. We cannot know what it meant twelve millennia ago to raise and bury monolithic representations of people, predators and prey. But this doesn't mean, to borrow a line from the British historian E. P. Thompson, that we should subject these ancient humans to 'the enormous condescension of posterity'[17] and assume that their sole achievement was, eventually, to become us.

Similarly, the foundational delusion this book is aimed against is that technology has immemorially driven our species along a certain road: that human history should be understood as an innovation-driven progression from past to present; and that a similar progression offers the only fit template for debating our future.

Understanding our exponential age

What follows when we reject the notion that innovation drives history along a preordained path? Among other things, attending more closely to the past's convolutions can help us better understand today's exponentially escalating complexities – and remind us that the range of possible futures we face is far wider than many might think, wish or fear.

Time in the human sense doesn't mean much when it comes to technology because, unlike something living, a tool doesn't itself

TECHNOLOGY IN DEEP TIME

struggle to survive or to pass on its pattern. Without its makers and maintainers, it is nothing. For all their ferocious sophistication, twenty-first-century technologies are like the earliest stone hand-axes in this respect. Without human manufacture, maintenance and refinement, they are merely matter. Untended, the fires of every factory and power plant will go out. Metal will rust, electricity ebb from wires, data decay. The fantasies of abandonment that haunt films and literature will fulfil themselves, unobserved.

No time passes for a technology unless it is used and adapted. If a human population deploys thousands upon thousands of identical farming tools in an identical way for thousands of years, that technology is frozen in stasis. To use an ancient tool is to enact a kind of time travel. And to forget how to use a tool – or for the infrastructures of knowledge and supply sustaining it to crumble – is to slip backwards in technological time, perhaps onto a different branch of its fractal possibilities.

This is rich fuel for the imagination. Speculative fiction is packed not only with possible futures, but also parallel pasts: steampunk empires, where the information age was built from brass gears and sweating boilers; neo-medieval dystopias, within which empiricism was extinguished by dogma. The human-made world is contingent in ways that constantly challenge us to rethink what it means to be human: to be tool-making creatures caught up in systems we created but did not choose.[18]

While we experience this history through much the same biological apparatus as our pre-technological ancestors, however, the world we've made has no such continuity. So far as technology is concerned, most of our planet's history saw no time passing whatsoever. Four billion years were less than the blink of an eye – while the last few centuries loom larger than all the rest of history.

There's an alluringly simple mathematical way of thinking about this. When it comes to combining things, increasing the number of components you're working with vastly increases the number of potential combinations. Three modules can be combined in six different ways, assuming each module is used once; four modules can be combined in twenty-four different ways; and by the time you reach ten modules, there are over three and a half million combinations. What this means is that, thanks to the fertile recombination of ever more technological possibilities, time and evolution are steadily speeding up from our creations' perspective. And the rate at which they're speeding up is itself increasing.

This has been most familiarly stated in the form of Moore's law, which began as an observation about the manufacture of transistors in a 1965 paper written by Intel's co-founder, Gordon Moore. With what turned out to be a remarkable mix of prescience and modesty, Moore noted that since the end of the 1950s:

> the complexity for minimum component costs has increased at a rate of roughly a factor of two per year . . . Certainly over the short term this rate can be expected to continue, if not to increase. Over the longer term, the rate of increase is a bit more uncertain, although there is no reason to believe it will not remain nearly constant for at least ten years.[19]

At the time Moore was writing, chips boasted around 64 transistors. By 1975, he suggested, there might be as many as 65,000 transistors per chip. This figure wasn't, in fact, surpassed by a commercial chip until 1979.[20] But the increasing complexity of computing showed no sign of slowing down, and the trend of doubling complexity every two years continued for the next half-century. In January 2023, Apple announced that its latest high-end commercial chip, the M2 Max, would feature over 60 *billion* transistors and be capable of performing over 13 *trillion* calculations per second.[21] Indeed, despite regular suggestions that transistor size and density are approaching their limits, the cost of computing performance itself continues to follow Moore's curve – as does the number of transistors in the world.[22]

This is the point at which technology starts to do strange things to time. Among the implications of these exponential increases, futurist thinkers such as Ray Kurzweil have argued that the next two years are likely to see as much progress in raw computing terms as the entire history of technology from the beginning of time to the present; and this is also likely to be true for the next two years, and the next, and the next. 'We won't experience 100 years of progress in the twenty-first century,' Kurzweil argued in his 2001 essay 'The Law of Accelerating Returns' – because 'it will be more like 20,000 years of progress (at today's rate).'[23]

You may have encountered versions of this analysis so often that it feels overfamiliar, or overstated. We can recapture something of its shock, however, by putting things slightly differently. From the perspective of certain technologies, humans have been getting exponentially slower every year for the last half-century. In the realm of software, there is more and more time available for adaptation and improvement. Outside it, every human second takes longer and longer to creep past. We – creatures of flesh and blood – are out of joint with our times in the most fundamental of senses.

Falteringly, we are beginning to face up to these facts. Consider one of the defining myths of our digital age, that of the Singularity: a technological point of no return beyond which, it's argued, the evolution of technology will reach a tipping point where self-design and self-improvement take over, cutting humanity permanently out of the loop. For Kurzweil, we have already reached the tipping point preceding this tipping point thanks to the creation of computation. The result? A near-future in which technology and divinity look remarkably similar. As he put it at the end of his 2001 essay:

Once a planet yields a technology creating species and that species creates computation (as has happened here on Earth), it is only a matter of a few centuries before its intelligence saturates the matter and energy in its vicinity, and it begins to expand outward at the speed of light or greater. It will then overcome gravity (through exquisite and vast technology) and other cosmological forces (or, to be fully accurate, will maneuver and control these forces) and create the Universe it wants. This is the goal of the Singularity.[24]

Is any of this likely? Like most myths, the least interesting thing we can do with this story is take it literally. Instead, its force lies in the expression of a truth we are already living: that the present power and influence of our technology have no precedents; and that its potential tragedy is the scale of the mismatch between the impact of our creations and our capacity to control them.

The most important lesson that follows from this, however, is precisely the opposite of Kurzweil's. Technology can neither dissolve our current challenges nor save us from ourselves. But it may, if we're able to embrace and interrogate it as an integral aspect of our humanity, allow us to inhabit this world in a new way: one informed by fresh forms of knowledge about ourselves and the systems sustaining our existence; one able to accept and embrace our evolutionary history, rather than building fantasies of escape from biology.

Can we deflect the path of technology's needs towards something like our own long-term interest, not to mention that of most other life on this planet? Not if we surrender to the seduction of thinking ourselves impotent or inconsequential – or of technology's future as a single, predetermined course. Like our creations, we are minute in individual terms but of vast consequence collectively. It took the Earth 4.5 billion years to produce a human population of 1 billion; another 120 years to produce 2 billion; then less than a century to reach the 8 billion humans currently alive, contemplating their future with all the tools of reason and wishfulness that evolution and innovation have bequeathed.

This is what existence looks like at the sharp end of history. Humanity is unique: uniquely responsible, uniquely capable, uniquely guilty. It is our technology that bears witness to and sustains this uniqueness; that blesses and curses us with a collective duty of self-invention. And it is our most ancient accomplishments – the stories we tell, the futures we dream – that best equip us to bend this path beyond present reckonings. We have less time than ever, and more that we can accomplish.

CHAPTER 2

Stone and fire:
The delusion of mastery

What images – if any – do you associate with human evolution? If you're anything like me, one of the first pictures the phrase may conjure in your mind's eye is a kind of procession. At one end is an ape-like creature, hairy and stooped. At the other is a recognizably modern (and quite possibly male) human, striding erect and hairless towards the edge of the frame. Between them are a number of progressively more upright and less ape-like ancestors, embodying humanity's ascent from its animal origins through primitive personhood to present glory.

There have been many versions of this image, but its prototype was called 'The Road to Homo Sapiens' and was created by the artist Rudolph Franz Zallinger for the 1965 book *Early Man*, part of a twenty-five-book series called the *Life Nature Library*.[1] Often referred to as 'The March of Progress', the image implies an evolutionary trajectory that's as straightforward as it is self-flattering: a linear, inexorable progression across millions of years towards humanity's present ascendancy.

Even in 1965, human evolution was known to be more complex than this. Indeed, the text and labels accompanying the original illustration make it clear that several of the figures were either contemporaries, evolutionary dead ends or out of sequence. But the procession's visual impact outstrips these qualifications – especially given the truth it undeniably contains, that we are the sole survivors of our evolutionary branch. There are no other surviving hominins.[2] All roads lead to *Homo sapiens*. QED.

One problem with hindsight is that, unless you're careful, it can fool you into assuming things couldn't have worked out any other way. In retrospect, Neanderthals were always going to die out; this politician was always going to lose, while that one was always going to win; the iPhone was always going to be a hit. The past becomes a story that makes sense in the light of present knowledge, in the process becoming utterly unlike the cloud of uncertainty and possibility that surrounds each unfolding moment. By turning history into tidy tales of cause and consequence, we risk losing touch with the most urgent lessons it can teach: that everything could easily have been different; that things will, over time, become more different than we can possibly imagine; and that causes, effects, purposes and progress are not features of the world, but of the stories we tell about it.

If you're looking for an image of our evolution that comes closer to the truth, try imagining not an orderly procession but an endless obstacle course: one shrouded in mist, predominantly gentle, but studded with terrible hazards. There is no finish line and no turning around, but there are islands of safety: terrains that allow respite, regrouping and reorganization; that may offer shelter for millennia before vanishing beneath fire or ice. Most importantly, there are no solo runners. Every species surges onward in gaggles of young and old, scouting and assisting and sacrificing. Sometimes, the groups help one another. Sometimes, they fight or mingle or splinter. Seen from the most distant of perspectives, they dash around their planet's pitfalls in swelling and shrinking tides, proliferating into new groups and types. And almost all eventually vanish.

If you look back between twenty and thirty million years, you'll see our ancestors first starting to pursue a distinct path from the rest of their primate class. Monkeys – smart, long-tailed climbers – spread across the Americas, Africa and Asia as early as forty million years ago. Around twenty-five million years ago, some monkeys began to exploit an evolutionary niche that allowed them to shed their tails, adopt a more upright posture and develop increased intelligence. At some unknowable point – one defined, like all divisions between species, by degree rather than any clear divide – these monkeys became what we now call apes.

A handful of species of ape remain scattered across the Earth today: gorillas, bonobos and chimps in central Africa; orangutans and gibbons in Southeast Asia. But their numbers are few, and their ability to survive present perils is in doubt. While monkeys remain widely distributed, apes are a tiny and endangered fraction of the planet's current species. There's one notable exception, of course: us. We are the apes who inherited the Earth. Yet we nearly weren't. And the story of our emergence turns out to entail not only near-extinction, but also a degree of commingling with our kin that marks our present isolation as wholly unrepresentative.

Fossils and DNA evidence suggest that chimpanzees and humans began to diverge from a common ancestor between six and eight million years ago, giving rise to the first creatures we would in due course call *hominins*: progressively less hairy, more bipedal and smarter apes. By around four million years ago, eastern Africa saw the first genus believed with some confidence to be our direct ancestor: *Australopithecus*, the 'southern ape'.[3]

Australopithecus ranged across much of sub-Saharan Africa. They didn't have significantly larger brains than chimps, and were around a third smaller than modern humans, but – as noted at the start of the previous chapter – they seem to have been crafting and using basic stone tools from more than three million years ago.[4] Inter-related species of australopiths thrived across a variety of habitats, from scrub and grassland to forest and lakeside, boasting an intriguing mix of ape- and human-like features. They had a vertical posture, big toes and stiff feet unable to grasp branches; but also long arms and curved digits. Males were almost double the size of females. Tool-usage, in other words, predated anything even approximating to humanity; a discovery that, alongside our growing awareness of other species' mental sophistication and diversity, has in recent years upended earlier accounts of tools as unique to our immediate ancestry.[5]

By around two million years ago, things had changed again. Three markedly different groups of hominins now coexisted in Africa.[6] *Australopithecus* was soon to become extinct in its original form, while several hominins of a new genus it had spawned – *Paranthropus* – were also dying out. But a new genus of hominin, *Homo*, had emerged in the form of (at least) two species: first *Homo habilis* and, slightly later, *Homo erectus*.[7] *Homo habilis*, 'handy man', was named for its apparent adeptness at identifying and shaping stone tools for butchering and skinning; while *Homo erectus*, 'upright man', was named in the 1890s based on the incorrect assumption that all previous hominins moved on all fours.

Both species likely represented a further increase in terms of intellectual skills and social sophistication. Only *Homo erectus*, however, seems to have managed two things that neither its ancestors nor its peers ever achieved. It bridged the divide between basic tools and more ambitious creations, for the first time bringing a truly technological culture into being. And it migrated beyond the African continent, engendering a global diaspora of hominins. Both of these undertakings helped ensure its survival. But they also entailed a series of extraordinary risks – ones that would, eventually, claim every descendant apart from us.

How tools became technologies

What's the difference between a technology and a tool? The previous chapter described technology as the entirety of the human-made artefacts that extend and amplify our grasp of the world. As it also noted, however, the emergence of technology entailed a distinction best thought of as a threshold: the point beyond which a culture of habitual tool-usage became one of evolving, combined and recombined artefacts.

When a capuchin monkey uses a stick to dig into a nest mound for eggs, the stick is an opportune extension of its agency.[8] We have long known that primates are smart in ways that echo our ingenuity; while, more recently, we have begun to take note of how many other species also make intelligent use of their environments' offerings, from crows and elephants to otters and dolphins. Indonesian octopuses are a personal favourite. They have been observed scavenging discarded coconut shells from the sea floor, emptying them of sand and mud, then using an elaborate 'stilt-walking' technique to transport the shells before pairing them up to protect their soft bodies.[9]

Seen in these terms, the earliest hominin tool-usage was impressive but not yet uniquely remarkable. Around the time *Homo erectus* appeared, however, things started to change. From 1.7 million years ago, simple stones with sharpened edges began to be replaced by so-called 'hand-axes', crafted in a standardized manner to match their wielders' grasp. What's known today as the Acheulean tool-making industry (it's named after an archaeological site at Saint-Acheul, near Amiens in France, where plentiful examples of such tools were found in 1859) existed from around 1.7 million to 100,000 years ago, making it by far history's most enduring manufacturing tradition. These personalized hand-axes were also the first objects truly to deserve the status of a technology.[10]

Over the vast period of time for which they were used – most likely for butchery, digging or carving – the production of Acheulean hand-axes progressed from the rough chipping of assorted stones to a specialized, skilled form of labour. Stone 'cores' were selected from suitable deposits, then sharpened to a remarkable consistency with the aid of multiple secondary tools. Eventually, the Acheulean tradition resulted in artefacts whose distinctive teardrop shapes perfectly matched the hands of their wielders: their symmetrical edges carved flake by flake with an expertise that would have taken years to master; their functionality married to beauty.

The crucial point here is that creating such tools demanded both the existence of skilled individuals able to pass on an accumulating body of knowledge and some generalized notion of what a hand-axe *ought* to be, independently of the particular material it was crafted from. To quote W. Brian Arthur's *The Nature of Technology* once again, 'new technologies are constructed mentally before they are constructed physically.'[11] The very possibility of a shared, taught and evolving technological culture entailed new forms of mapping between mental and physical realms.

Once again, recent research continues to push back the frontiers of intellectual and cultural sophistication. Today's dating of the Acheulean industry to 1.7 million years ago is several hundred thousand years earlier than was once thought to be the case, with further sites continuing to be found across the African continent.[12] Evidence of advanced hand-axe 'workshops' has now been found from more than 1.2 million years ago, demonstrating the existence of a technical and problem-solving culture able to concentrate its labour in particular locations. As the authors of a 2023 study investigating a site in modern-day Ethiopia put it, far from merely 'coping' with environmental conditions, these hominins both planned around the seasonal floods that deposited the obsidian they carved and 'creatively solved through convergent thinking technological problems such as effectively detaching and shaping large flakes of the unusually brittle and cutting volcanic glass.'[13]

The Acheulean tradition continues to provoke speculation (even the verdict that its earlier forms deserve the status of a technology is not unanimous).[14] How and why did it endure for so long? Did it serve aesthetic, ceremonial or status-related purposes – or play a part in courtship, a theme explored by the delightfully named 'sexy hand-axe theory'?[15] Whatever the truth, it signalled the emergence of two entwined traits possessed to a unique degree by our ancestors: *imagination* and *mental time travel*. And these in turn offer some fundamental insights into technology's prerequisites and nature.

Imagination is implicit in the standardized form of technologized tools. Somehow, ancient hominins reached the point where they could look at a hunk of stone and see within it both the tool it might become and the purposes it might serve. What, though, does the enigmatic phrase 'mental time travel' describe? I've borrowed it from the evolutionary psychologist Thomas Suddendorf, who has written extensively about our ancestors' relationship with time; and, in particular, about how rich temporal experiences were instrumental in the emergence of culture, technology and language.

As Suddendorf notes, modern humans make little neurological distinction between past, present and future, or indeed between memory and imagination.[16] No other creature is capable of thinking outside the present moment with anything like our sophistication. Somehow, at around the time of technology's emergence, hominins' experiences of the world started shifting towards an open-ended and ongoing project of speculation, reconstruction and collaboration. Their minds became capable of both *simulating* and *sharing* accounts of potential world-states. And their technologies were reflections and extensions of these faculties: iterated investigations of what might prove useful; snatches of memory, observation and imagination mobilized into new forms. Remarkably, incrementally, they began to remake the external world in the light of their inner lives.

The last of the hominins

Fittingly enough for so curious and accomplished a tribe, the first technological traditions coincided with hominins' earliest global migrations. Within a few hundred thousand years of its emergence, *Homo erectus* had moved far beyond Africa, expanding first into southern Eurasia, then into Southeast Asia and Indonesia. It would survive in Indonesia until as recently as 150,000 years ago. And its flourishing over such a time and distance saw the emergence of a variety of descendants almost unthinkable from our present vantage: a planet populated, albeit sparsely, by interrelated yet profoundly different proto-human intelligences.[17]

Looking back 1.5 million years, it's tempting to see *Homo erectus'* global migration and cultural sophistication as proof that evolution's most difficult work was done the moment tools began to become technologies. Earth boasted its first technologized, imaginative, sociable hunters, and this surely meant that – sooner or later – their descendants' triumph was assured. Indeed, a version of this tale has long been told about our origins: of humanity's preordained ascendancy. 'Let us make man in our image', declares the God of Genesis, 'and let them have dominion over the fish of the sea, and over the fowl of the air, and over the cattle, and over all the earth'.[18]

Once again, however, growing knowledge of the past suggests a more uncertain path. Between 1.7 and 1.2 million years ago, *erectus* and other hominin populations continued to mingle, breed and develop. Among other things, this period saw the emergence of what was probably our own most direct ancestor, a species sometimes known as *Homo heidelbergensis* (a designation that may in future be replaced by *Homo bodoensis*), which evolved from those *Homo erectus* populations that had remained in Africa.[19] More human- than ape-like in their features, with a meat-rich diet suggesting sophisticated hunting tactics, the little we know about this species fits a narrative of growing intelligence and social complexity. Yet recent genetic research also suggests that around 1.2 million years ago, the entire global population of hominins crashed to fewer than 20,000 individuals.[20]

Whatever its cause (climate change is one contender), the result was a genus that clung precariously to existence for hundreds of thousands of years. That *Homo* survived at all is a tribute to its tenacity. That it nearly didn't spells out its vulnerability, and the degree to which the earliest forms of technological collaboration and ingenuity were far from guaranteed routes to thriving. Hominins were, and are, unprecedentedly smart in unprecedented ways. But time and evolution don't care about such details. The only guarantee is that every successive obstacle must either be met by adequate adaptations or – as has proved the case for 99.9 per cent of the species that have ever existed – succumbed to in the form of extinction.

By around half a million years ago, some members of *Homo heidelbergensis* or *bodoensis* had taken advantage of a warmer era between ice ages to migrate into Europe from Africa, where they would grow apart from their African kin into species including Denisovans and Neanderthals. After the ice returned, this migration ceased for several hundred thousand years, during which time the first biologically modern humans emerged in Africa: our species, *Homo sapiens*, who start to appear in the fossil record around 300,000 years ago.[21] We had arrived, more or less. But it wasn't until around 100,000 years ago that we began our first major migrations out of the African continent: last and youngest of the hominins.[22]

Homo sapiens entered a world already home to scattered cousins with whom we continued to mingle and interbreed for tens of thousands of years. As recently as 75,000 years ago, the ghost of extinction shivered into view again: a supervolcanic eruption at the site of Sumatra's present-day Lake Toba which, by some estimates, caused a global reduction of *sapiens* to fewer than 10,000 breeding pairs. We survived, as did Neanderthals, Denisovans, *Homo floresiensis* (the so-called 'hobbits' of the island of Flores, Indonesia), and other as yet imperfectly understood lineages. But it was – depending upon your preferred interpretation – a close-run thing.[23]

Our relationships with other hominins endured for most of our existence, and have left deep traces. All modern non-African humans carry fragments of the Neanderthal genome, while some Denisovan DNA survives in Oceania and parts of Southeast Asia. In aggregate, all non-African modern humans are estimated to owe between 2 per cent and 7 per cent of their genes to archaic humans.[24] Just 40,000 years ago, Neanderthals still walked southern Europe.

These species were not only smart but also culturally and technologically accomplished. The genetic evidence irrefutably tells us that we raised generations of offspring together – something that, authors like the linguist Sverker Johansson have argued, in turn demonstrates a deep common capacity for culture and language.[25] Yet, by the time the last ice age ended, 12,000 years ago, *sapiens* found itself alone at the end of the evolutionary line. Every single one of our closest relatives had fallen by the wayside. What happened to them – and why didn't it happen to us?

It's plausible, of course, that *we* are part of the answer to this question; that over time we out-competed, out-fought and displaced our kin. But geography, genetics and climatological data suggest another overarching factor: failure to adapt to our planet's ever-shifting states.[26] Indeed, perhaps the greater wonder is not so much that they *didn't* survive as that we *did*, for they and we possessed many of the same talents and aptitudes. We committed to the same exchanges: of specialism for adaptability; of toughness and instinct for technological culture. And through almost our entire existence, this know-how was at best a dubiously beneficial asset.

A central lesson of our survival, in other words, is precisely the opposite of the one evolution is often assumed to teach: that humanity is the pinnacle of a natural and inevitable order, and that our brilliance guarantees our thriving. Hence this chapter's subtitle: the *delusion of mastery*, a reference to the belief that we have somehow stepped outside the logic governing all other life. There is enough truth in this delusion to make it seductive. In the larger scheme of things, however, it is both profoundly self-deceived and a recipe for catastrophe.

Entering an age of fire

At this point, let's take several steps back. What do we modern humans look like in the context of evolution's obstacle course? Compared to every species dashing madly, blindly alongside us, we are – for want of a better word – *cheats*. They have no idea what lies ahead or came before. If their environment becomes less hospitable, they must either move, evolve, adapt their behaviours or die. These are the rules of the game, and almost everything that has ever lived has no more power to change them than it does to weaken gravity or switch off the sun. We, however, do; because we alone have learned to harness the forces and phenomena that constitute our world. And perhaps the most fundamental force of all, whose mastery fuelled both our survival and our eventual ascendancy, is also the most elemental: fire.

Consider what happens when, today, I wake up and realize that it's going to be unseasonably cool. The south of England doesn't get particularly cold, even in winter, but if it's going to lurk below freezing for a few days I'll set our central heating system to run continuously, governed by a thermostat. If I look through the small window in the front panel of our big gas boiler, I see flames dance: blue peaks fading to white. This is as close as daily life brings me to tending a fire. The boiler does its thing while beneath the street outside my house metal pipes deliver a ceaseless supply of fuel. It's an exchange that, in its simultaneous ease and planetary impact, embodies all that technology has given across the millennia – and may yet take away.

Like everything else we burn, the energy in this fuel was first captured from the sun by living organisms: by plants and the animals that fed upon them, in the case of natural gas, many millions of years ago. To burn such fuel is to swap old life for new heat. And without life, no such exchange would ever have taken place on Earth. Towards the start of her 2020 book *Transcendence* – to which the analysis of this chapter as a whole is indebted – the author Gaia Vince points out something at once fundamental and astonishing about the relationship between life and fire:

> For the first billion or so years of Earth's history, there were no fires because there was nothing to burn and no oxygen to burn it. It took the evolution of photosynthetic bacteria, followed much later by the growth of the world's first forests, for there to be the ingredients of fire. Life itself had to generate the environmental conditions of its own destruction.[27]

Ever since, fire has been a force of nature in the blindest and most literal sense. Ecosystems have evolved to thrive on periodic burnings. Ice ages have seen oceans rise and fall, deserts shrink and spread. Sporadic cataclysms have brought immolations and extinctions, throughout which life has adapted and survived – and continued to fuel the possibility of its own ending. Yet between one and two million years ago, most likely in the hands of *Homo erectus*, fire began for the first time in history to be controlled rather than simply encountered: to become, falteringly and incrementally, harnessed to proto-human culture.

The opportunistic use of fire must have been known long before it was deliberately controlled: the capture of naturally occurring flames for heat and protection. Indeed, hominins aren't even the only creatures to make use of wild fire. As Australia's Aboriginal people have long known, so-called 'fire hawks' can intentionally spread wildfires by transporting burning sticks in their beaks or talons (they do this in order to drive out prey).[28] Hominins, however, were unique in learning not only how to capture fire from wild sources but also how to keep it alive in hearths – then put it to work.

What did it mean and entail to achieve this? Fire and heat are such effortless features of our world today that we tend to project this ease backwards: to depict our ancient ancestors bashing together a couple of flints, watching sparks fall onto tinder and twigs, then whooping delightedly at the resulting flames. Although its mechanics are simple enough, however, the business of safely setting, starting and tending fires is far from straightforward. It demands carefully gathered and crafted components, cooperative labour, a comprehension of the differing roles of tinder, kindling and fuel – and thus involves not only intellect but also a society at least as sophisticated as that accompanying standardized stone tools. It was, not to put too fine a point on it, immensely effortful and counterintuitive to arrange a survival strategy around artificial warmth.

For obvious reasons, fire leaves behind a patchier fossil record than stone, but there are signs of burnt materials in sites containing other evidence of hominin activity from as long ago as 1.5 million years; and of possible hearths from more than half a million years ago. From 400,000 years ago the trace evidence becomes denser, perhaps marking an improvement in fire management; a theory supported by the emergence of 'hafting' techniques in tool-making, meaning the attachment of stone tips to wooden shafts with twine and glue, an exercise likely to have required heat treatment.[29]

What did fire mean to those who made and tended it? Above all, it meant life. Stone tools allowed us to hunt, cut, scrape and craft skins and branches into new forms. But it was flames that truly transformed our relationship with the material world. As Vince puts it:

> Whereas our earliest human ancestors had bedded down in tree nests for safety, fire protected their descendants from predators and the cold, allowing them to sleep in open savannahs. Fire culture was adapting our species' habitat for their survival; as fire made our world safer, we altered the environmental selection pressures acting on our genes.[30]

Fire fuelled many of the first feedback loops between biological and technological evolution, intensifying and accelerating both of these in the process. It created warmth and safety in the absence of hairy bodies and woodland habitats. It allowed us to range further and more freely; to craft finer tools; to increase through cookery the calories and nutrients available from food. For the first time in planetary history, energy was being domesticated. For the first time, a species was knowingly rather than instinctually harnessing a source of energy greater than itself.

Sapiens went far further, and faster, than *erectus* had ever done in the harnessing of flames. And it's in the degree of *sapiens*' reliance upon artificial heat that perhaps the best clue to our survival lies: in the ways our vulnerability and weakness may have become a source of strength. *Homo erectus* had long vanished from Africa by the time *sapiens* appeared, and its isolated communities in Southeast Asia had also become extinct before our ancestors' arrival there. For a brief time, however, *sapiens* shared western Eurasia with the last of our ancient kin, *Homo neanderthalensis*.

Based on the best available evidence, Vince imagines the Neanderthals' final generations settled near caves at the base of what we now call the Rock of Gibraltar: cooking, eating together, threading eagle talons onto necklaces, carving intricate patterns into stone by firelight. 'These are people with rich interior lives,' she writes,

> with time to think and create art. Deep inside in the cave, past the little sleeping chambers with their individual protective fires, there is a special nook containing a deliberately carved rock engraving: a crosshatch of parallel lines. Its symbolic meaning will be lost in the befuddling layers of time . . .[31]

By the time *sapiens* arrived, the Neanderthals seem already to have been in terminal decline, most likely thanks to a combination of climate change, inbreeding and disease. They didn't know their species was dying. Or did they? Neanderthals were smart, creative and richly communicative; human enough to interbreed with us. So far as climate and fate were concerned, however, they may have been insufficiently fragile. Well adapted for the cold of interglacial Europe, with larger brains and more powerful bodies than our own, they never travelled further from their origins than central Asia. They may simply have required too many calories and wandered too little to survive time's accumulated obstacles – although they lived long enough to gift us a fragment of their genome.

Was *sapiens* driven by a more restless, relentless curiosity? Whatever the reasons, Neanderthals joined this planet's other hominins in oblivion while we embarked upon migrations to its farthest reaches. And we carried with us the gift of domesticated energy: a

Promethean spark that would one day bake clay, smelt metal, fuel furnaces and transform the future of life on Earth.[32]

The bargain struck by *Homo sapiens* began with fuel and flames, and this is also where it may end. We have labelled ourselves 'wise humans' in tribute to the fact that language, culture and learning make us unique; that they have driven us faster and further than any other species. Yet the story of our exponentially accelerating impact upon this planet is most starkly told in terms of energy. In geological terms, cold rather than heat has ruled recent history. Over 800,000 of the last 900,000 years have been icy, with the entirety of what we call civilization contained within a dozen postglacial millennia. It was climate that most likely sealed other hominins' fates, and a return to ice is overdue. Thanks to humanity's incineration of millions of years of captured light within the space of a few centuries, however, we are instead lurching in the other direction: towards unprecedented heat at unprecedented speed.

There are several names for the era of human influence on the Earth – the *Holocene*, the *Anthropocene* – but one of the most recent speaks to energy itself. In a 2015 essay for *Aeon* magazine, the environmental historian Stephen J. Pyne suggested that we are now living in the *Pyrocene*: an age of fires and burnings. As I write these words, in late 2023, atmospheric levels of carbon dioxide are 420 parts per million, an increase of almost 50 per cent on pre-industrial levels. The last time the atmosphere held anything like this amount of carbon dioxide was three million years ago, when global temperatures were three degrees centigrade warmer than the present and sea levels fifty feet higher – and the genus *Homo* didn't exist.[33]

So far as the future is concerned, all bets are off when it comes to climate. The natural glacial cycles that defined our evolution may be a thing of the past, but what will replace them is unlikely to be gentle. As Pyne puts it:

Something seems to have broken the rhythms [of natural glacial and interglacial periods]. That something is us. Or more usefully, among all the assorted ecological wobbles and biotic swerves that humans affect, the sapients negotiated a pact with fire. We created conditions that favoured more fire, and together we have so reworked the planet that we now have remade biotas, begun melting most of the relic ice, turned the atmosphere into a crock pot and the oceans into acid vats, and are sparking a sixth great extinction . . . fire has become as much a cause and consequence as ice was before. We're entering a Fire Age.[34]

In environmental terms, the first decades of the twenty-first century offer a foretaste of what an age of fire will be like: one characterized, unlike those of ice, by speed and volatility; by oceans and an atmosphere stirred into ferocious instability; by ecological losses to rival our planet's previous mass extinctions.

Yet, Pyne notes, there is also hopefulness – of a kind – within fire itself. Fire is our evolutionary companion, our foundational technology. And it remains ours to harness; to grasp in all its systematic consequences, if only we can acknowledge these in time.

Between ice and fire, ice is the more terrible. It obliterates what it mounds over; it crushes and drives off life. By contrast, fire is a creation of the living world: life gave it oxygen and fuel and, with people, ignition. Its fundamental chemistry is a biochemistry that takes apart what photosynthesis puts together. It cannot exist without life. We can manipulate fire, directly and indirectly. We can't ice. We survive ice by leaving. We survive fire by living with it. If at times it seems our worst enemy, it is also our best friend. We can't thrive without it.[35]

To return to this chapter's subtitle, the story we most often tell about ourselves is one of mastery: of landscape, of other species, of ourselves. There is truth in this, up to a point. We have bent the Earth to our will. As its systems start to pass beyond certain tipping points, however, we are discovering that it no longer yields. Perhaps it is time for us to learn another lesson: that mastery is no longer a fit aspiration for our species.

CHAPTER 3

Love and learning:
The delusion of brutality

In stones and flames, we have looked at the technologies that distinguished hominins from other apes alongside the intellectual capacities they relied upon. Yet this still tells only half the story: that which made us remarkable, but not the drives that lay beneath it. What sentiments and longings burned in our ancestors? What can we know about the texture of their lives as they adapted and endured?

One of the most important answers to both these questions has little to do with ingenuity and everything to do with a word less commonly used in the context of technology: *love*. Indeed, one of the most remarkable features of our species is the breadth and depth of the love that exists not only between parents and their children, but also between kin, non-kin and generations; and the unexpected ways in which this has proved both a prerequisite for technological cultures and their driving force.

To begin with, consider a central paradox of human thriving. We are the only hominins to have survived the last million years. However transiently, we dominate this planet. Yet we suffer from an immense vulnerability compared to every similar creature: the duration and fragility of human childhood.

A human child cannot hold up its own head for the first few

months of its life; cannot move itself around for half a year; cannot walk before a year; does not start to gain adult strength or resilience for over a decade; does not finish sexual maturation for around a decade and a half; and continues to experience substantial growth and development in the brain's prefrontal cortex for two and a half decades. No other mammal takes so long to grow into independence. None acquires so many fundamental behaviours and capabilities via nurture rather than nature. Human offspring are staggeringly burdensome from the moment of their delivery – which both requires assistance and represents a huge hazard to mothers – through much of their long reproductive immaturity.[1]

These are the costs. But what about the benefits? In her 2009 book *The Philosophical Baby*, the philosopher and psychologist Alison Gopnik sums them in a single word: *change*. More than any other creature that has ever existed, Gopnik writes,

> human beings are able to change. We change the world around us, other people, and ourselves. Children, and childhood, help explain how we change. And the fact that we change explains why children are the way they are – and even why childhood exists at all.[2]

Change and childhood are, in evolutionary terms, two sides of the same coin. It is the remarkable neurological plasticity that goes hand in hand with childhood that has allowed humanity to develop its cultural, technological and intellectual skills. And childhood entails this plasticity precisely because such skills have proved so advantageous.

Indeed, technological ingenuity was entwined with the biology of childhood well before *Homo sapiens* even existed. There's mounting evidence, for example, that as long as four million years ago *Australopithecus* infants were far larger in relation to mothers than those of other apes; and that this was bound up with bipedalism, brain growth and the emergence of 'carrying' behaviours later supported by 'external wombs' in the form of baby-carrying slings.[3]

Dating such technologies is inherently speculative (being crafted from plant or animal matter, they leave no archaeological remains), but in his 2010 book *The Artificial Ape*, the archaeologist Timothy Taylor goes so far as to date baby slings to almost two million years ago, crediting them as enablers of the rapid expansion of brain size that distinguished our subsequent lineage. By helping our ancestors to transport and protect smart-but-weak offspring, it became feasible for the skulls – and thus the brains – of these offspring to keep on growing long after birth. The width of the birth canal no longer limited the brain's potential. By contrast, all other primates' offspring must actively cling onto their hairy parents, and are thus born in a relatively developed state, limiting the postpartum neurological development they can undergo. A human degree of dependency is simply impossible for any other species.

Even if you don't accept the full extent of Taylor's chronology, technological practices were undeniably integral to our ancient ancestors' extended childhoods: the construction and maintenance of fires, shelters, clothes, weapons, carriers, containers. And these developments in turn permitted ever longer periods of infant vulnerability, neuroplasticity and learning. Not only is there no such thing as a human existence free from tools and culture; we have also laboured across hundreds of thousands of years to permit ourselves ever *greater* flexibility and dependency, both upon one another and our creations. As Taylor puts it:

> We have never been wholly natural creatures, and we have evolved to be increasingly artificial. Even should we want it, escape from technology is no longer possible. It may in fact be that technology has escaped us . . . Either we crash, or we continue our artificial ascent.[4]

Where does love fit into this picture? Most importantly, the vulnerability of human childhood *demands* an adulthood that is its mirror image. Fully grown, we are as helpful as our children are helpless; as competent as they are demanding; as empathetically collaborative as they are dependent. How could we be otherwise? As commentators like the American anthropologist and primatologist Sarah Blaffer Hrdy have noted, human child-raising demands networks of collaboration reaching far beyond the nuclear family: forms of communal care and labour without which our other accomplishments would be impossible. We are empathetic, hypersocial creatures – and the gulf between us and other primates is defined as much by these attributes as by intellect. Here's how Hrdy puts it in her 2009 book *Mothers and Others*:

> Once acquired, the habit of comparing humans with other primates is hard to shake . . . Descriptions of missing digits, ripped ears, and the occasional castration are scattered throughout the field accounts of langur and red colobus monkeys, of Madagascar lemurs, and of our own close relatives among the Great Apes. Even among famously peaceful bonobos, a type of chimpanzee so rare and difficult to access in the wild that most observations are from zoos, veterinarians sometimes have to be called in following altercations to stitch on a scrotum or penis. This is not to say that humans don't display similar propensities . . . But compared with our nearest ape relations, humans are more adept at forestalling outright mayhem. Our first impulse is usually to get along . . . From a tender age and without special training, modern humans identify with the plights of others and, without being asked, volunteer to help and share, even with strangers. In these respects, our line of apes is in a class by itself.[5]

This innate interest in others' minds encompasses not only children's extraordinary capacities for learning, but also adults' extraordinary capacities for teaching: an endeavour we coordinate across networks including parents, siblings, aunts, uncles, grandparents and non-kin. Indeed, humans are alone among primates in according special status to post-reproductive adults, whose knowledge, experience and assistance make them invaluable as carers and cultural custodians. No other apes experience anything like the human menopause, which creates cohorts of post-reproductive females with decades left to live (among all other species, orcas perhaps come the closest to us in this respect).[6] And this hyper-collaborative custodianship is eminently fit for purpose. To care for culture – to teach, iterate and innovate – is as valuable to our species as reproduction itself. Indeed, in its potential to enrich and safeguard countless future lives, it may be still more precious.

As Hrdy emphasizes, none of this is to deny that human nature and history feature terrible instances of violence, intolerance and cruelty; or that tribal and familial passions can breed atrocities. Mutual care, however, remains by definition more fundamental than conflict; while narratives that downplay this risk presenting a distorted version of history. As she puts it:

> textbooks in fields like evolutionary psychology devote far more space to aggression, or to how men and women competed for or appealed to mates, than they do to how much early humans shared with one another to jointly rear offspring. Even when human hypersociality is noted, explanations tend to emphasize between-group competition rather than how difficult it was to ensure the survival to breeding age of costly, slow-maturing children. Yet . . . without shared care and provisioning, all that inter- and infragroup strategizing and strife would have been – evolutionarily speaking – just so many grunts and contortions signifying nothing.[7]

Across ice ages and extinctions, our ancestors managed to adapt to the challenges of their environment – and to adapt this environment to them. And the evolutionary strategies enabling this were driven not by cool intellect, but by the depth and multiplicity of their capacities for love: for children, parents, kin and comrades; for knowledge, skill and beauty.[8]

The word 'love' doesn't often feature in textbooks discussing evolution and survival, perhaps because it can seem unscientific or even sentimental: an unnecessary projection of private emotion onto empirical facts. Yet to speak of such things in the context of our shared history is one of the most important ways we can remind ourselves that conquest, aggression and mastery are less than half the human story.[9] Similarly, although it can seem an enigma demanding explanation, childhood is only a burden when seen through adult eyes. From the species perspective, the wonder is not so much the effort adults put in as the immense return children deliver on even the meagrest investment – and how far an emphasis on individual experience neglects what's most remarkable about us. As Gopnik concludes:

> If we focus on adult abilities, long-term planning, swift and automatic execution . . . then babies and young children will indeed look pretty pathetic. But if we focus on our distinctive capacities for change, especially imagination and learning, then it's the adults who look slow.[10]

This is a logic that applies far beyond reproduction. No matter *who*, *what* or *how* you love, it is the collective project of doing so that bears our species forward. Our vulnerability – our enduring, empathetic interdependency – is also our greatest strength.

What should we take from all this? Among other things, it's profoundly important in terms of the stories we prioritize and the lessons we believe history can teach. As someone who regularly attends twenty-first-century technology conferences and visits the headquarters of tech companies, I'm repeatedly struck not just by the relative absence of women in many such venues, but also by the dearth of serious discussion of such topics as compassion, child-raising and nurture.[11] There are, of course, exceptions. All too often, however, 'technology' is conceived by those at its cutting edge in the crudest of Darwinian terms: as an arms race that only the fittest and most aggressive will survive.

All of this is almost amusingly unfaithful not only to the struggles and priorities that brought our species to this point, but also to Darwin's own work.[12] Hence this chapter's subtitle, *the delusion of brutality*, which refers to the misreading of evolutionary science as a tale of unending conflict: a vision of history and technological development cast solely in terms of conquest, not to mention of human nature as ineradicably steeped in violence. We can, and must, do better. And this begins with an acknowledgement of what it actually means for each of us to grow from our brilliant, vulnerable beginnings into members of a living culture.

The R&D department of the human species

Although we can only guess at the sheer variety of cultures that have flourished and coexisted across the millennia, we can state with a high degree of confidence that love, nurture and learning were their supreme survival strategies. Indeed, our collective capacity for innovation and technological recombination only makes sense when seen in these terms: as a form of learned, imitated and improvised iteration reliant upon the openness of each generation to both old knowledge and fresh explorations. Here's Gopnik on the intergenerational dynamics of culture and technology:

> There's a kind of evolutionary division of labour between children and adults. Children are the R&D department of the human species – the blue-sky guys, the brainstormers. Adults are production and marketing. They make the discoveries, we implement them.[13]

Consider my own children. As I type these words, my son and daughter are aged six and four years old respectively. They're running around the garden, screaming and laughing and pretending to be various animals – a cat, a dog, a dragon – while I try to focus on the crawl of type across my screen. In what ways are, and aren't, they like the first of our ancient ancestors to tread the ground we live upon?

Britain was sporadically visited by archaic humans from around 900,000 years ago, but *Homo sapiens* doesn't seem to have reached it until a mere 40,000 years ago. And *sapiens* didn't settle here until later still, most likely thanks to ice ages' regular scouring of this archipelago. Given that even 40,000 years isn't long in terms of mammalian evolution, this means that almost everything that makes my children different to the first young *sapiens* to play here relates to culture and technology.[14]

The clothes my children are wearing; the (relative) cleanness of their skins; the language they're speaking; the knowledge they carry in their minds: all of these things are specific to the present. Yet, if I had somehow been handed two newborn late Palaeolithic babies to adopt, they would happily have mastered everything that my children have done. They too would be modern humans – because modernity is a description not of our biology or nature, but of the environment within which we are raised.

There are some evolutionary differences, of course, many of which relate to how culture and innovation have shifted survival's parameters. My son has blue eyes, a feature of humanity for only the last six to ten thousand years.[15] My children and I can digest lactose, a tolerance developed over approximately the same period thanks to the domestication of milk-producing livestock.[16] It's also likely that my children have slightly smaller and more efficient brains than their ancestors, smaller jaws, and that they may not develop their upper and lower wisdom teeth.

More remarkable than such differences, however, are the similarities. What we now call the Palaeolithic era spans almost the entirety of prehistoric human development, from the emergence of the first proto-humans to the point at which, 12,000 years ago, the most recent glacial period ended and humanity began its accelerating global growth. We *sapiens* are a notably homogenous species, with a narrower gene pool than the one shared by hominins at the time of Neanderthals and Denisovans. But we more than compensate for this via the staggering fecundity of our intellectual lives.

The opportunities and burdens of this legacy are a heady mix. As many commentators have noted, one challenge of modern times is that our biological inheritance can seem wildly out of kilter with present conditions. Leda Cosmides and John Tooby, two of the founders of the field of evolutionary psychology, put it bluntly:

> Our modern skulls house a stone age mind. The key to understanding how the modern mind works is to realize that its circuits were not designed to solve the day-to-day problems of a modern [citizen] – they were designed to solve the day-to-day problems of our hunter-gatherer ancestors.[17]

It is certainly true that our unaided instincts can be outmatched by the speed, scale and complexity of some twenty-first-century challenges. But there's also a potentially dangerous form of condescension wrapped up in this framing – and in the related belief that what our 'stone age brains' most need is rational interventions to tame their flaws. For while there is much to admire in exhortations to reasoned self-transformation, it's easy to over-value the insights of behavioural and cognitive research while underestimating the subtlety and flexibility of our stone age 'circuits'. And that's before you get to the determinism implied by any analogy between a human mind and a circuit board.

What did it mean for countless previous generations to suffer, love and hope while bringing all that we now have into being? For the Palaeolithic archaeologist April Nowell, one of the most notable absences in many current accounts of ancient life concerns children themselves: their actions and experiences; their place in the world. As she put it in February 2023:

> We forget that the adults of the Paleolithic were also mothers, fathers, aunts, uncles and grandparents who had to make space for the little ones around them. In fact, children in the deep past may have taken up significantly more space than they do today: in prehistoric societies, children under 15 accounted for around half of the world population. Today, they're around a quarter.[18]

Children are partly absent from accounts of the ancient past – at least as anything other than appendages to adults – because their small, fragile bones are hard to find and preserve. But researchers have also proved prone to projecting backwards their own assumptions, and to depicting ancient children as mere adults-in-waiting: cherished, playful, socially uninvolved. The accuracy of such assumptions may seem difficult to test. Yet, Nowell notes, plenty of evidence can in fact be found, if sought with sufficient care. The stone flakes produced by novice Palaeolithic tool-makers, for example, actually outnumber those made by expert adults in archaeological sites around the world.

Archaeologists can recognise the work of a novice because people learning to produce stone tools make similar kinds of mistakes. To make, or 'knap', a stone tool, you need a piece of material such as flint or obsidian, known as a 'core', and a tool to hit it with, known as a 'hammerstone'. The goal is to remove flakes from the stone core and produce a shaped blade or some other kind of tool . . . novices, who were often children or adolescents, would sometimes hit too far towards the middle of a core, and each unskilled hit would leave material traces of their futile and increasingly frustrated attempts at flake removal. At other times, evidence shows that they got the angle right but hit too hard (or not hard enough) . . .[19]

Not every novice was a child, of course; but nor was every expert an adult. Like the few tribal societies that still exist today, Palaeolithic humans would necessarily have learned essential skills from an early age: hunting, butchery, digging and foraging; creating tools and containers and slings. There's also considerable evidence that children were active participants in communal bodies of practical and symbolic knowledge. In almost every chamber of France's eight-kilometre Rouffignac cave system, for example, where a remarkable range of finger paintings from 13,000 years ago are preserved, methodologies developed over the last two decades have made it clear that we are looking at the work of children's as well as adults' hands. As Cambridge archaeologist Jess Cooney, a member of the pioneering archaeologist Leslie Van Gelder's team, put it:

Some of the children's [paintings] are high up on the walls and on the ceilings, so they must have been held up to make them or have been sitting on someone's shoulders. We have found marks by children aged between three- and seven-years-old – and we have been able to identify four individual children by matching up their marks . . . The most prolific of the children who made flutings was aged around five – and we are almost certain the child in question was a girl. Interestingly of the four children we know at least two are girls. One cavern is so rich in flutings made by children that it suggests it was a special space for them, but whether for play or ritual is impossible to tell.[20]

Emotionally, intellectually and communally, these were not 'primitive' people. Their minds were different from ours – a difference that testifies to the profound transformations wrought by culture and technology. But their achievements also emphasize a common ground of adaptability, care and wonder. As Gopnik notes, children are anything but passive receptacles for knowledge. Their 'stone age brains' are sites of wild and radical innovation, eager to imitate and exceed their role models: to embrace and extend the treasure store of their society's accomplishments.

Our species is indeed biased and cognitively constrained. But treating our brilliant vulnerability as mere biological baggage – an outmoded inheritance we should aim to transcend – risks conceiving history as a hierarchy within which the 'modern' is inexorably superior to an imagined past, not to mention any insufficiently modern aspects of the present. Moreover, it implicitly abases the present at the future's feet, casting us as imperfect creatures awaiting deliverance from frailty: irrational animals whose only hope is to submit to technologized optimization.

Such a framing not only confuses wisdom with progress but also risks equating progress with a narrowly technologized vision of rationality. Our potentials for species-wide self-knowledge are precious, as is the growth of scientific and technical knowledge. But this doesn't mean that certain twenty-first-century models of what is, and isn't, reasonable should automatically count for more than the biology that bore us through thousands of generations. There are better ways of framing our gifts and our vulnerabilities. And articulating these begins with the realization that mutual care, compassion and empathy are more enduring adaptations to deep time than dreams of self-perfection.

Cooperation is our species' superpower

In her 2021 book *The Social Instinct*, the evolutionary and behavioural scientist Nichola Raihani makes a large claim about the significance of human hyper-sociality as an evolutionary strategy: that we can only hope fully to account for its benefits (and burdens) if we look far beyond hominins, primates and even mammals for precedents.

> Cooperation is our species' superpower, the reason that humans managed not just to survive but to thrive in almost every habitat on Earth . . . It is tempting to focus on our closest living relatives – the great apes, and especially chimpanzees and bonobos – but this approach is blinkered. Social behaviours that have the distinct whiff of humanness about them are often absent in apes and monkeys but do appear in much more distant connections. Ants and meerkats teach, for example, but chimps don't. Pinyon jays will share their stuff, but bonobos won't.[21]

Our supreme commitment to cooperation, Raihani suggests, is both 'the key to solving the massive global problems we now face' and an ability 'that might also be our eventual downfall' – because, given the freewheeling force of our imaginations, it simultaneously enables immense collaborative feats and invidious social comparisons. In particular, this double-edged potential is manifested in the unique strength of humans' interest in *relative* rather than absolute rewards; and in our talent for ourselves defining the parameters by which success, failure and purpose are measured.

Another way of putting this is that humans care almost as deeply about others' experiences and opinions as their own; and that almost all our judgements are sociable and contextual rather than absolute. Other primates demonstrably show disappointment if they receive a reward that fails to meet expectations: that is, they are sensitive to *insufficiency*.[22] Beyond this, however, they are generally happy to accept what they see as a good deal regardless of what their fellows receive. By contrast, even young children are innately inclined to compare their own payoffs with others' and to reject *unfair* allocations: that is, to apply an inherently relative standard.

In one 2014 study, which investigated the attitudes of children across a range of countries to sharing, participants as young as four were given the chance to accept or reject a range of different offers. One of the experiment's scenarios entailed two children being offered a tray of sweets each, with one child allocated the role of 'decider'. The decider could choose whether both participants got what was in their tray, or whether both trays were tipped away, meaning nobody got anything. The sweets were sometimes allocated fairly and sometimes unfairly, with one child having four sweets in their tray versus just one in the other. While equal deals were accepted around 80 per cent of the time, even four-year-old 'deciders' accepted poor deals only 60 per cent of the time – and older children less than half the time – despite the fact that doing so meant they received no sweets rather than some.[23]

These results chime with decades of research suggesting that humans' unique capacity for cooperation is underpinned by a strong aversion to unfair outcomes; and that this aversion is sufficiently strong that we are willing to pay a cost in order to 'punish' outcomes perceived as unfair.[24] This particular experiment, however, also suggested some additional complexities. In some instances, the 'decider' faced an unfair allocation in their *favour*: four sweets in their own tray versus one in the other. Up to the age of seven, this advantageously unequal distribution was accepted around 80 per cent of the time. Eight- and nine-year-olds, however, rejected this form of unfairness around half the time. They chose to have nothing rather than enjoy the fruits of a rigged game.

The stakes, of course, were low – and the interactions face-to-face – but this is still a striking result. As the study's authors noted, younger children seemed to be motivated by the desire to punish perceived unfairness *against* their interests, but were broadly happy to take advantage when it counted in their favour. Older children, however, showed an interest in fairness as a generalizable principle: as something that *ought* to govern interactions whether or not it happened to suit them.

What can we take from this? We shouldn't need experimental evidence to tell us that people have the potential both to be discontented with even the most staggering surfeit, and to prize fairness even at great personal cost. When it comes to the connection between these behavioural poles, however, perhaps the most striking point is also the most fundamental: that how we *frame* questions of value and priority matters more than any apparently absolute standards; and that, as children, we generally grow from judging fairness purely in terms of personal payoffs to judging it as an impersonal principle, its details conditioned by biology and culture alike.

It is, in other words, 'natural' for us to think, judge and care outside the present moment: to look to one another, and to the words and tools we have collectively brought into being, in order to understand our minds and needs.[25] And there is within this a vertiginous freedom. Evolutionary psychology can teach us a great deal about the common ground we share with our ancestors. But we must be careful that its lessons don't blind us to the breadth of the strategies we have discovered – and may yet discover – for transcending our limitations: behaviours and innovations that echo everything from ants' hive-minds to the mimetic intellect of birds; from cetaceans' intergenerational care to the fractal forms of coral cities or termite mounds.

Similarly, the crudity of the *delusion of brutality* lies in its myopic focus on conflict in the context of a species that has immemorially sought and found, through childhood's vulnerability and adulthood's collective care, fresh forms of thriving. Every word and tool we use is a communal project – and so too are our desires and judgements. They speak through us. They enact the norms and expectations of a particular time and place, and these norms are at once sociable, communal and artificial. As the philosopher Richard Rorty puts it in the first chapter of his 1989 book *Contingency, Irony, and Solidarity*:

> The world does not speak. Only we do. The world can, once we have programmed ourselves with a language, cause us to hold beliefs. But it cannot propose a language for us to speak. Only other human beings can do that.[26]

To be creatures of language and imagination is to call beliefs, judgements and values into being; to submit our own and others' minds to a constant form of collective scrutiny. But it is also to possess minds that are more than merely biological: that live and dream beyond the body's confines.

CHAPTER 4

Extended minds:
The delusion of 'it' and 'us'

'Where does the mind stop and the rest of the world begin?' This is the question with which the philosophers Andy Clark and David Chalmers begin their 1998 paper 'The Extended Mind'; and its argument has become a staple of discussions of technology's relationships with thought and identity.[1]

The extended mind hypothesis, as the paper's core claim has come to be known, offers a tantalizingly counter-intuitive answer to its opening question. Most people, when asked to specify the boundary between their minds and the world, naturally gesture towards their own bodies. After all, the human mind is demonstrably the product of a brain in a body; and this in turn suggests that everything beyond the skin is demonstrably *not* part of this mind. We may observe or interact with the outside world, but in each case we are simply deciding (in the privacy of our own minds) to do particular things in relation to something external.

Clark and Chalmers, however, propose a very different model, based on 'the active role of the environment in driving cognitive processes.' To the extent that external aspects of the world play an active role in our thinking, they suggest, our minds are constantly 'extended' beyond our bodies – and our thoughts and actions cannot be understood without reference to these extensions. We should thus accept that certain aspects of the external world are *literally* parts of our minds: that we integrate them so naturally and entirely into aspects of our thinking that such thinking can neither be properly described nor, in some cases, even exist without them.

Here's an illustrative scenario. Imagine for a moment that I'm visiting Athens, a city I don't know very well, and want to find a cafe where I can get a decent cup of coffee and something to eat. What do I do? Probably, I get out my mobile phone and search online for recommendations. I'm likely to do this in conjunction with a mapping app, so that I can find somewhere conveniently located. After I've selected what looks like a good venue, I'm likely to use this app to plan my route. While following this route, I may read a few more reviews, or look at a menu in advance, or exchange a few messages with friends and family. Once I've arrived, I may use my phone to automatically translate the menu in front of me, or to facilitate a conversation with a waiter. I'll probably pay using my phone too. And if I've had a particularly good (or bad) experience, I may write an online review to help others work out whether they want to visit this particular place.

To what extent, in the situation above, does it make sense to describe my thoughts and actions only in terms of what's happening inside my head? Or – to put the principle behind this question more generally – am I *the same person* with and without my mobile phone in my hand? Certainly, things would play out very differently if I found myself in Athens without a phone. I wouldn't be able to look up the details of cafes online, or their locations, or read reviews. I wouldn't have a GPS and mapping app. I wouldn't have instant access to messages from friends and family in my pocket. Would all this alter my mind as well as my options?

One way of exploring this question is to consider what I might do *instead* of using my phone. If I were to find myself in such a scenario, among my first actions (apart from buying a cheap replacement phone) would be to get hold of a city map, a guidebook and a basic Greek phrasebook. I would, in other words, seek to replicate via older technologies the same advantages an internet-connected phone offers – because it's hugely valuable for me to be able to navigate a strange city, to understand some of the local language and to know what's worth visiting.

None of this would be possible if I lacked these things. And until these things were invented, none of these possibilities existed in anything like their present form. Accurate maps are an extraordinary technology, achieved over many centuries: one which transformed humanity's ability to store and share detailed geographical information. Comprehensive, accurate dictionaries and guidebooks are similarly powerful, and dependent in turn upon centuries of development in our collective capacity to store and transmit knowledge. Before this, travel and exploration relied upon word of mouth and human intermediaries – translators, fixers, guides – as well as upon various forms of custom and hospitality.

While its fundamentals still entailed leveraging others' expertise for mutual advantage, in other words, the world was very different – and so were we. Just a century ago, the very notion of strolling through Athens receiving real-time information, directions and bilingual translations from a hand-sized hunk of metal and glass would have seemed like a delusion on the far side of improbability. And the fact that I've grown up in a world where I do such things on a daily basis means that aspects of my thoughts and experiences are profoundly different to those of people who didn't grow up in such a world; a difference that applies whether or not I have a phone to hand.

None of this is to say that any particular technological developments are self-evidently good things any more than they are inevitable. Indeed, the abstraction of activities such as asking for directions into the realm of technology has undoubtedly entailed losses as well as gains. Depending upon your preferences, you may sometimes *choose* to go 'offline' in order to experience a place free from technological mediation; to use a physical rather than digital guidebook or map; or to use an augmented reality app to overlay virtual enhancements on your vision. All of which activities, in their own ways, only emphasize the significance of different technologies to cognition.[2]

Importantly, the extended mind hypothesis is a theory of *mind* rather than of *consciousness*. As Chalmers has clarified in subsequent writings, he and Clark weren't suggesting that an individual's consciousness literally extends into their environment.[3] Instead, they were describing how elements of the mind's 'background' like memories and beliefs can be outsourced. Seen in these terms, it's clear that there's nothing inherently new about people using both their bodies and their environment to perform mental functions. As Chalmers notes in his 2022 book *Reality+*, some cognitive aids are as fundamental as it gets:

> The first time someone counted on their fingers, the process of counting was being partly offloaded from brain to body. The first time someone used an abacus, the work of calculation was being offloaded from the brain to a tool. The first time someone wrote something down for later use, the work of memory was being offloaded from brain to written symbols.[4]

Digital technologies represent, in this sense, not so much a break with the past as a radical intensification of something ancient; a kind of collective unconscious in their pooling of billions of minds' worth of memories, ideas, claims and capabilities.[5] Acknowledging this means rejecting the delusion that there's a hard divide between 'it' and 'us': that technology is simply something *out there* in the world that we can choose to use, reject or resist.

As the rest of this chapter explores, such a view is not only inaccurate but also profoundly misguided. Strange though it may seem, some technologies are indispensable aspects of our minds, shaping not only *how* but also *what*, *whether* and *when* we think. And this makes the assumptions and possibilities embedded within them of pressing importance.

Writing, rhetoric and the extended mind

What is the practical significance of the extended mind hypothesis? Most fundamentally, it emphasizes the impossibility of describing or accounting for our thoughts without addressing the apparatuses of enhancement, outsourcing and abstraction surrounding every human life. Consider my own children once again. At six and a half years old, my son is only two years older than my daughter. Yet between them lies a divide that defines our species' leap into modernity. This is because my son can read and write while my daughter, as yet, cannot.

Literacy has been with us for six thousand years, give or take: far too recent to entail innate biological elements in the way that spoken language does. Across the first few years of their lives, all healthy human children learn to speak the language of their particular culture. They do so as inexorably and informally as they learn to walk. Unless children are carefully educated across a number of years, however, written language means nothing to them. Every one of us is born into prehistory and must be taught a way out.

What happens when a child is taught to read? Neurologically, what's going on is the repurposing of an apparatus that evolved to perform quite different tasks: recognizing faces, objects, depths, lines; classifying similarities and differences; abstracting meanings and possibilities. Culturally, what's happening is the learned association of a series of human-made artefacts – letters or characters – with the phonemes composing speech. This is the plasticity of the human mind writ large: a colossally labour-intensive training process that, after years of effort, grants almost every resident of the twenty-first century access to the recorded realm of words and meanings.

Neuroplasticity is at root a description of certain biological facts. Literate brains are markedly different to illiterate brains, with notable changes in behaviour and connectivity; in particular, part of the left occipitotemporal cortex develops what's known as a 'visual word-form system'.[6] Literate minds are similarly distinctive. From our perception of space and time to our memories, from our beliefs to the structures of our societies, written and recorded culture has transformed what it means to be human.[7] Or, to put it another way, literacy epitomizes how a species whose evolution depended upon manipulating the physical world through technology has also learned to manipulate symbolic and informational worlds – and, by doing so, has further unshackled itself from the constraints of flesh and time.

Like the distinct language of mathematics, the words you're reading right now are a technology whose lineage can be traced back to markings made in ancient Mesopotamia six millennia ago, where a system for tallying goods gradually came to be associated with the spoken monosyllables of ancient Sumerian. Alphabets are young enough that the form of some letters has persisted through the majority of writing's existence: O, which likely began as the Egyptian hieroglyph for the eye; B, which (when turned on its side with the straight line on top) was a hieroglyph for shelter, the letter's loops representing a door and a window; A, which, if you flip it upside down, still resembles the horned ox it represented in ancient Semitic languages. In fact, every single modern alphabet ultimately descends – via the ancient Phoenician and Greek trading empires – from Semitic speakers' adaptations of hieroglyphs near the start of the second millennium BCE.[8]

Writing has been with us for an evolutionary eyeblink. Yet through almost all of this time it has remained a minority pursuit. As recently as 1940, according to data from the OECD and UNESCO, around six in ten adults across the world remained illiterate.[9] Today, that figure is less than one in ten, marking one of the most central of the twentieth century's cultural transitions: the opening up, for the first time in history, of written records to a majority of our species. Still more remarkably, the last few decades have seen via the spread of the internet and mobile devices the rise of mass *participation* in written and recorded culture. Cumulatively, these events embody perhaps the greatest enhancement of human scope since writing itself arose. But they have also seen a new intensification of old debates around the power and consequences of technologies' extension of our mental lives.

Throughout literacy's history, some people have suggested that writing represents too great a power to be widely taught; and that, by divorcing words from their speakers, it risks destabilizing the very prospect of knowledge and understanding. This might seem a bizarrely retrograde line of thought in the twenty-first century. Yet we can find among these ancient arguments some startlingly enduring insights. There is, for example, a famous moment towards the end of Plato's philosophical dialogue *Phaedrus* – written around 370 BCE – in which the character of Socrates discusses with his interlocutor, the eponymous Phaedrus, what he calls 'the propriety and impropriety of writing'. Here are the words Plato puts into Socrates's mouth:

> I cannot help feeling, Phaedrus, that writing is unfortunately like painting; for the creations of the painter have the attitude of life, and yet if you ask them a question they preserve a solemn

silence . . . And when they have been once written down they are tumbled about anywhere among those who may or may not understand them, and know not to whom they should reply, to whom not: and, if they are maltreated or abused, they have no parent to protect them; and they cannot protect or defend themselves.[10]

Writing, unlike the living practice of philosophical dialogue, self-evidently cannot speak for itself or understand what is asked of it. And this makes it dangerous. As the character of Socrates puts it in a previous parable about writing's invention, to record things is to risk both diminishing the reader's mental faculties and encouraging the ignorant performance of knowledge:

> This discovery of yours will create forgetfulness in the learners' souls, because they will not use their memories; they will trust to the external written characters and not remember of themselves . . . you give your disciples not truth, but only the semblance of truth; they will be hearers of many things and will have learned nothing.[11]

There's an obvious irony to the fact that Plato's words only survive because, unlike those of the actual Socrates, they were written down. But this doesn't make them wrong. If you go online today and look up Plato's *Phaedrus*, you'll find not only dozens of translations of the original text, but also thousands of pre-digested quotes, views and reviews, waiting for you to copy and paste them into a document or presentation. If you're writing a book about technology and want to give it a flavour of classical erudition, Plato's critique of written words is ready to be used and abused, just as he feared: an effortless parroting of ancient wisdom. Indeed, perhaps the commonest contemporary use *Phaedrus* is put to is one that explicitly ignores its actual argument: as a way of suggesting that new technologies have always provoked anxiety and thus, given that books are self-evidently a good thing, that these anxieties have always been unfounded.

Ironies aside, Plato was writing at a remarkable moment in history. The Mediterranean world was drifting slowly from an oral to a written culture. But this culture was young, its norms uncertain, its gifts and losses yet to be determined. Both Socrates and another yet-to-be-born great teacher, Jesus of Nazareth, remained entirely within the oral tradition. But Plato – like the disciples who preserved Jesus' story – was determined to record and expand upon his master's teachings. And among the first forms used for the resulting written records were simulacra of human voices and interactions: dialogues and letters, eternally captured on the page.

All of this still matters two millennia later. As the author James Gleick notes in his 2011 book *The Information*, the ambivalence at the heart of *Phaedrus* embodies a remarkable alertness to information technology's mingled risks and potentials:

> [Plato] witnessed writing's rising dominion; he asserted its force and feared its lifelessness. The writer-philosopher embodied a paradox. The same paradox was destined to reappear in different guises, each technology of information bringing its own powers and its own fears.[12]

Plato feared that his readers would perform his ideas in the absence of understanding, like bad actors stumbling their way through a script. Yet still he expressed himself in writing, in the process producing works whose depth and intricacy no oral tradition could match; whose power lay in their creation of a realm with its own rules, potentials and pitfalls. Twenty-four centuries later, in a world suffused with the novelty of mass participation in written and recorded culture, this informational realm still bristles with the same problems of trust, provenance and persuasion.

Of all these themes, one that exercised Plato more than any other remains perhaps the most urgent: the relationship between what people *wish* to believe and what's actually going on. In the dialogue *Gorgias* – written around 390 BCE, two decades before *Phaedrus* – Plato addressed a practice potentially inimical to truth-telling and truth-seeking: rhetoric, the ancient art of persuasive speech. Like Phaedrus and most of the other main characters in Plato's dialogues, Gorgias was a real person. Born in modern-day Sicily in 483 BCE, just over half a century before Plato, he was one of the most renowned orators of his age, famed for bringing rhetoric to mainland Greece and for the brilliance of his public performances. The first half of Plato's *Gorgias*, however, makes a ferocious case against rhetoric's persuasive arts, with the character of Socrates casting it alongside cookery and cosmetics as a form of mere people-pleasing:

> The whole of which rhetoric is a part is not an art at all, but the habit of a bold and ready wit, which knows how to manage mankind: this habit I sum up under the word 'flattery'; and it appears to me to have many other parts, one of which is cookery, which may seem to be an art, but, as I maintain, is only an experience or routine and not an art.[13]

The words Plato puts into Socrates's mouth here have a bitter additional sting. The actual Socrates had been executed (for impiety and corrupting the young) in 399 BCE following a trial that – as was the nature of Athenian democracy at the time – entailed rival rhetorical appeals being made to a council of five hundred men. Such a trial is alluded to in the *Gorgias* as resembling a competition between a doctor and a cook, judged by a jury of children; that is, as a performance within which tasty untruths all too easily win favour against bitter medicine:

> If the physician and the cook had to enter into a competition in which children were the judges, or men who had no more sense than children, as to which of them best understands the goodness or badness of food, the physician would be starved to death.[14]

Rhetoric and democracy deeply discomfited Plato, and for related reasons: people are only too easily seduced by sugar-coated untruths and flattery's mechanical routines. In both historical Athens and the dialogue bearing his name, Gorgias embodied a tempting intellectual path. He profited handsomely by teaching wealthy Athenians to wield words as weapons. But by doing so, Plato suggests, he made verbal skill a purely instrumental business: a handmaiden to society's structures of power and self-aggrandizement.

At its worst, rhetoric is a form of flattery indifferent to truth, its sweetly seductive claims aiming at persuasion by any means. Hence Plato's advocacy in his most famous work, *The Republic* (c. 375 BCE), of enlightened leadership by male and female philosopher-guardians – and his outlawing of the 'wrong' sort of poetry. In Plato's ideal state, art's function is educative, and this demands the suppression of passionately misleading words. As the character of Socrates argues at the start of Book X:

> I do not mind saying to you, that all poetical imitations are ruinous to the understanding of the hearers, and that the knowledge of their true nature is the only antidote to them.[15]

Was Plato wrong? In one sense, he was gravely mistaken about language and art alike. Censorious, dictatorial, elitist, his vision aligns truth with an austere ideal of intellectual authority that few today would endorse. In another sense, however, his advocacy of philosophical dialogue and the poetic elegance of his prose undercut such dogmatism. Although the character of Socrates often seems to speak for his author, nowhere does Plato actually appear in his own work. Rather, he depicts conversation – at its best – as a joint and equal endeavour, aimed at interrogating assumptions in a world where consensus should never simply be imposed.

It's perhaps for these reasons that the questions Plato raised continue to resonate so remarkably. What duties do information technologies, not to mention those who make and maintain them, have to their audience? If – to echo *The Republic*'s most famous parable – most people live their lives chained in a cave watching shadows play on the wall, does someone who knows about the real world beyond the cave have an obligation to free them? Should that person shrug and let things be; offer the prisoners a choice between pleasing illusions and painful truths; or forcibly bring them into the light?[16]

Coupled systems: merging our minds with the world

The questions above have a particular force in the light of the extended mind hypothesis – as does resisting the tendency to dismiss them merely as matters of personal preference. If technologies are not only useful tools or delightful entertainments, but also aspects of the minds reliant upon them, the question of the duties accompanying this reliance becomes an urgent one. We live, in Plato's terms, in a world suffused with records and rhetoric: with ubiquitous mediated manipulations. And this in turn suggests one of modern mental life's central tensions: the degree to which meaningful choices may rely upon the information environments framing and incentivizing them.

While the essence of this tension is ancient, addressing it means paying close attention to present particulars. Just as the development of books and maps brought new kinds of cognition into being, so have today's technologies: types of cognition that can't be adequately described by saying that a mind largely unchanged since the Pleistocene is doing new things. To do so would be to confuse a generic brain in a generic skull with the altogether more astonishing result of a particular human being born into a particular time and place; growing and learning within a particular cultural and technological context; and engaging constantly with other minds through the shared apparatus of words, ideas and tools. In one sense, 'I' am a brain strolling around inside a body, and this brain and body aren't too different to those of my ancient ancestors. But in another, more significant sense, 'I' am a mind that spills far beyond this body along countless contingent lines of action, interaction and comprehension.

Similarly – to return to my initial example – having a mobile phone doesn't just affect my identity, in the sense of how I perceive and present myself. As I use and come to rely upon it, my phone becomes a kind of intellectual prosthesis that not only enables but also *enacts* aspects of cognition. Take it away, and I am altered in non-trivial ways. Install new apps, unveil new features, and I am non-trivially augmented. And I am by definition unaware of many of the most significant aspects of this augmentation until absence or alteration forces me to confront them.

Here, Clark and Chalmers's reasoning can help us to draw out some important distinctions. Noting the 'tendency of human reasoners to lean heavily on environmental supports', they suggest that there's a fundamental difference between using tools purely to bring about a particular state of affairs (picking up food with a fork) and using tools to perform part or all of a cognitive process (rotating a shape on a computer screen to work out whether it will fit into a particular slot). This second category of usage demands 'epistemic credit' – that is, credit for being part of a mental undertaking that would otherwise have been impossible.

Perhaps most significantly, the concept of epistemic credit invites us to think about people and various technologies as 'coupled systems' whose properties can only be understood by considering what such systems achieve as a whole. The ubiquity of high-quality cameras on modern devices has, for example, been accompanied by a suite of behavioural innovations that can only be fully explained as products of the coupled system 'a-skilled-person-plus-a-phone-with-camera'. When I take a picture of a menu in a foreign language, then use an app to instantaneously overlay an English translation onto this image, I'm deploying a capability that it makes little sense to attribute independently either to me or my device. When a river of traffic flows along a freeway, its dynamics can only be understood as the collective actions of countless cars-plus-humans. When I look something up online, this seemingly simple act entails the mediation of my interests and desires through a vast, distant index of billions upon billions of web pages.

All of the above entails a transformative mix of losses and gains. Tom-plus-his-phone is a system capable of perfect visual recall and endless connectivity. By contrast, Tom-minus-his-phone is notably poor at remembering things that a couple of decades ago he knew by the dozen: friends' and family's phone numbers, street names, email addresses. So long as my phone is online, charged and to hand, I can grasp and interpret the world in profoundly new ways. But once I'm habituated to these possibilities, the absence of my phone is no longer neutral. Indeed, as anyone who's found themselves with a flat battery at a crucial juncture will know, it's felt as the lack of a significant cognitive and social capability.[17]

How should we make sense of all this? Towards the end of their original paper, Clark and Chalmers pose some questions that are at least as significant as their opening provocation, but that have received considerably less attention:

> If the thesis is accepted, how far should we go . . . Is my cognitive state somehow spread across the Internet? . . . What about socially extended cognition? Could my mental states be partly constituted by the states of other thinkers?[18]

Once again, looking at the lives and minds of children can be clarifying. Take my daughter, aged four, as she navigates each new day. She can already think about and understand the world in ways available to no other animal. But the evolutionary cost of these talents means she's constantly, eagerly dependent upon a host of humans and human-made objects to make her world safe and comprehensible. The food she eats; the water she drinks; the bed she sleeps in; the clothes she wears; the shelter and warmth of her home: her family and community not only provide all these, but are also constantly in the business of describing and discussing them.

My daughter's comprehension of the world is, in other words, a shared, ongoing project: one in which her mind's steady extension is central to identity and thriving. And the world of human-made artefacts and ideas surrounding her is not a neutral backdrop to her self-directed development. Every aspect of it, from the new words she acquires each day to the stuffed dinosaur she clutches in bed each night, is a richly experienced amalgam of interactions and opportunities – not to mention a compendium of century upon century of human ingenuity. To offer a final quote from Alison Gopnik, humans are unique in inhabiting an environment that is primarily born from our imaginations; that is itself the child of countless minds:

If I look around at the ordinary things in front of me as I write this – the electric lamp, the right-angle-constructed table, the brightly glazed symmetrical ceramic cup, the glowing computer screen – almost nothing resembles anything I would have seen in the Pleistocene. All of these objects were once imaginary – they are things that human beings themselves have created. And I myself, a woman cognitive scientist writing about the philosophy of children, could not have existed in the Pleistocene either. I am also a creation of the human imagination, and so are you.[19]

The further we move through history, the larger the human imagination and its works loom; and the more our minds are interwoven with artefacts embodying numberless legacies, experiments and iterations. This is one of the most obvious ways technological modernity can be said to alienate us from both other creatures and our own origins. At the same time, however, these ancestors remain right beside us in the newness of each child: in their adaptivity and fierce desire to learn; their deep interest in other minds; their playfulness and lack of presumption.

In my experience, one of the most humbling aspects of having children is being asked to explain 'why' the world is the way it is – and constantly failing to give satisfactory answers. Why do people have to work and earn money? Why do some people have so much, and some so little? Why do Mummy and Daddy love their phones so much? To be faced by a child's questions is to re-encounter history as contingency: as a series of steps that produced one of an infinite number of possible presents, and that holds the potential for an equal range of futures. And this means it's also a chance to relearn our collective capacity for challenge and change: to ask how things came to be the way they are, and what it might mean to remake them.

Am I part of my daughter's mind? Certainly, other people are non-optional when it comes to all kinds of action and understanding on her part. Outside of her immediate sensory impressions, my wife and I – and the friends, family and community ranged around us – are the primary interpreters and facilitators of a modernity that cannot be comprehended unaided, either by her or by anyone else. She may choose to reject, ignore or dispute countless aspects of her civilization. She will, I hope, live to see changes I can scarcely imagine. But there's no sense in which any of this can happen outside of a particular context and perspective. As we all must, she is busy stitching together a self from her unique experience of common ground: from the books and toys and songs, the tantrums and joys, of life within our home's walls.

Embodiment and extra-neural resources

Even if you accept its premises, there can seem something alienating about the extended mind hypothesis: an implicit hankering after human augmentation and biomechanical experimentation that, at the very least, isn't for everyone. The question of where a human mind ends and begins, however – and of the ethical imperatives bound up in the human-made world – becomes a little less strange if you think about it in bodily terms.

We are social animals, not abstract entities, and it's on this level that the most fundamental aspects of our morality and mental lives play out. To be safe and cherished is to experience something profoundly important to identity and development. To enjoy good health and liberty is to live in a different world from someone who suffers chronic illness or captivity. Privilege and inequality are, first and foremost, matters of the body. Human hearts and minds are most often changed not by abstract reason but by bodily experience: by matters that move us on a creaturely basis.

Indeed, the staggering plasticity of human learning and cooperation only fully makes sense in the light of consciousness's embodied nature. We all think and feel differently depending upon where we are and who we are with; upon the contexts, systems, habits and communities we occupy. And we do so even when we imagine what's going on is a purely intellectual business. As the behavioural economist Daniel Kahneman emphasizes in his 2011 book *Thinking, Fast and Slow*, 'cognition is embodied; you think with your body, not only with your brain' – a point he introduces by inviting his readers to look closely at two words printed next to one another:

bananas / vomit

These words are, of course, no more than a dozen letters on a page. Yet the very act of closely attending to them is likely to induce a host of involuntary physiological responses associated with disgust. Look at them again. What happens when you do so? Whether you notice it or not, your heart rate has probably increased slightly; your sweat glands activated; certain memories and associations started to stir. Alongside this, you'll have become more likely to notice and react to further words and ideas associated with *bananas* and *vomit*. And you may feel slightly less interested in eating.

'The essential feature of this complex set of mental events', Kahneman observes, 'is coherence. Each element is connected, and each supports and strengthens the others.'[20] An essentially arbitrary and under-specified stimulus – two words, nothing more – triggers a miniature cascade of associations embodying a deeply rooted form of sense-making. Why? Because words aren't just neutral vehicles for information any more than a home is just a dry space for sleeping in. Powerful, complex associations invariably flow through us on the basis of even the simplest prompts. We weave coherence out of the world around us, constantly, because this 'sticking together' of disparate ideas and sensory impressions is among our minds' most fundamental activities: one that entails constant feedback between brain, body and world.

Consider children and childhood one last time. As every twenty-first-century parent knows, children's relationships with technology are fuel for countless dubiously valid analyses of tech addiction, dependency and diminishment. Almost unmentioned in many of these debates, however, is the other side of the coin: the strength of the evidence around our need for such things as movement, touch and stimulation; for a sensorily rich range of activities and apprehensions; for good food, good company, rest and recuperation. 'Technology' isn't necessarily the enemy of any of these things. But neglecting them is most assuredly a recipe for diminishing our minds and identities: for needlessly narrowing the scope of the human.[21]

In fact, as the author Annie Murphy Paul points out in her 2021 book *The Extended Mind: The Power of Thinking Outside the Brain*, it's precisely because our minds are so *unlike* digital technology that they benefit from its enhancement. We think best, she suggests, when 'we think with our bodies, our spaces, and our relationships' – something that some technologies' tendency to treat us as disembodied can efface, as can equating human thought and feeling with information-processing.

Above all, what Paul terms the 'centrality of extra-neural resources to our thinking processes' is an indication of how misguided it is to liken the brain to some kind of central processing unit set splendidly atop the body's machinery – and how ignoring our embodied, sensory natures is one of the surest ways of creating technologies likely to diminish rather than enhance our potentials. 'Our devices can and do extend our minds,' Paul writes:

> but not always; sometimes they lead us to think less intelligently, as anyone who's been distracted by clickbait or misled by a GPS

system can tell you . . . Too often, those who design today's computers and smartphones have forgotten that users inhabit biological bodies, occupy physical spaces, and interact with other human beings.[22]

This echoes a final point in Clark and Chalmers's paper that feels at least as prescient as its premises. 'It may be,' they note in their concluding paragraph,

> that in some cases interfering with someone's environment will have the same moral significance as interfering with their person.[23]

To a degree that may have surprised even Clark and Chalmers, the dynamics of information systems have become a battleground for competing values – and competing claims about human nature, freedom and thriving. On 27 June 2016, the United Nations General Assembly adopted a non-binding resolution titled 'The promotion, protection and enjoyment of human rights on the Internet' which specifically addressed governments' denial of internet access to citizens as a form of coercion. The UN declared itself:

> deeply concerned by all human rights violations and abuses committed against persons for exercising their human rights and fundamental freedoms on the Internet . . . Deeply concerned also by measures aiming to or that intentionally prevent or disrupt access to or dissemination of information online.

And it went on to affirm:

> the same rights that people have offline must also be protected online, in particular freedom of expression, which is applicable regardless of frontiers and through any media of one's choice.[24]

As the second half of this book explores in greater depth, this direct identification of online freedom with other common goods echoes a fundamental facet of our age: that information environments increasingly mould not only our beliefs, interactions and opportunities, but also *how* we think and *who* we believe ourselves to be; and that distinctions such as online/offline are increasingly irrelevant when it comes to rights, identities and opportunities.

What's to be done? There are large themes here that invite us to explore how our minds are extended not only individually but also collectively by our creations; and the degree to which we outsource to automated systems not only physical and mental labour, but also aspects of our emotional, social and ethical selves. Paul's perspective is, however, a fine place to start, alongside a proposed approach that's as much pragmatic as it is hopeful. 'We extend beyond our limits,' she argues,

> not by revving our brains like a machine or bulking them up like a muscle – but by strewing our world with rich materials, and by weaving them into our thoughts.[25]

Collectively, incrementally, our minds can only grow towards their full potential by challenging the delusional dichotomy between 'it' and 'us' – and by embracing the degree to which our thoughts are never wholly our own.

CHAPTER 5

Consciousness as controlled hallucination:
The delusion of literal-mindedness

What is it about our minds that makes them so readily able to adapt, extend themselves and embrace the world? The answer lies in the intricacy of our access to reality – and in the ways that human perceptions and understandings constantly leap beyond the literal.

When I was young, I was fascinated by an optical illusion you've probably encountered yourself: a diagram in which two identical parallel lines are made to appear different lengths by the addition of two short arrow-like fins at each of their ends, pointing inwards in the case of one line and outwards in the case of the other.

What was fascinating was the persistence of the illusion in the face of decisive knowledge that it wasn't real. The error I made when I first saw the diagram – that I was looking at two lines of different lengths – was swiftly and easily corrected. But my perception of what I was looking at wasn't so easily swayed; and this was at once strange and marvellous.

Even though I knew that the parallel lines were the same length, and even though I repeatedly verified this with the ruler from my school pencil case, I couldn't make myself *see* this fact. No matter how long I stared, the line with the outward-facing arrows always appeared longer than the other. Unless, that was, I covered up both sets of fins with my fingers, at which point the lines went back to being nothing more than identical marks on a piece of paper.

I didn't know it at the time, but I was looking at a version of a diagram first described in 1889 by the German sociologist Franz Carl Müller-Lyer, illustrating what he called an '*Optische Urtheilstäuschungen*': an 'optical judgement-deception'.[1] The deception takes the form of a paradox: that the lines both are and aren't the same. They are clearly and verifiably identical on the page. At the same time, they are equally clearly and verifiably perceived as different lengths by human observers; and these perceptions can be manipulated in predictable ways.

Plenty of research has been conducted exploring Müller-Lyer's examples in the 130-odd years since his original paper. Adjusting the length and angle of the fins alters the strength of the illusion (the optimal fin/shaft ratio for maximum distortion is, apparently, around 36 per cent, while the optimal angle for the fins lies between 40 and 60 degrees).[2] Different cultures also appear to experience the illusion to differing degrees. But while there's no definitive agreement as to how such illusions work, what they *do* definitively tell us is that there's a gap between our perceptions and the way things actually are – and that knowing this gap exists both does and doesn't allow us to control it.

On the one hand, we literally cannot see some things accurately, no matter how hard we try. On the other hand, we can gain a remarkable level of insight into both the nature of reality and the limitations of perception, so long as we approach them with a sufficiently open mind. It's an intriguingly analogous situation to our relationships with technology. We do not and cannot choose the world we are born into – and nor can we shrug off the momentum of its systems or societies. Yet we can, if we are careful, enter into an informed negotiation with these things: one that acknowledges our mutual limitations while seeking a direction of travel aligned with hope and human thriving.

Consciousness as controlled hallucination

In his 2021 book *Being You*, the neuroscientist and philosopher Anil Seth begins a step-by-step analysis of consciousness by setting out the biological basis of perception. Every single one of our experiences, he emphasizes, ultimately depends upon electrochemical activity inside a brain sealed within the skull's darkness.

> There's no light, no sound, no anything – it's completely dark and utterly silent. When trying to form perceptions, all the brain has to go on is a constant barrage of electrical signals which are only indirectly related to things out there in the world, whatever they may be. These sensory inputs don't come with labels attached ('I'm from a cup of coffee', 'I'm from a tree'). They don't even arrive with labels announcing their modality – whether they are visual, auditory, sensations of touch, or from less familiar modalities such as thermoception (sense of temperature) or proprioception (sense of body position).[3]

As Seth notes, it's dangerously easy to assume on the basis of these facts that incoming signals directly 'encode' information which, via an exquisitely complex form of biological computation, the brain directly 'decodes' into sounds, images, tastes, smells and textures. Within each of us, according to this view, there sits a 'self' somehow attending to sensory inputs which reflect reality in a more or less transparent manner – as if we were watching the world via neurological CCTV. It's an explanation that can seem self-evident, partly because screens and computation are metaphors as central to our era as valves and boilers were to the nineteenth century. Yet, as our physiological and neurological understanding of the brain has improved, it has become increasingly clear that the actual processes underpinning perception look little like this.

For a start, the 'bottom up' view of an inner self monitoring impartial sensory inputs gets things precisely the wrong way around. So far as we can tell, perception is inextricably bound up with prediction: with a 'top down' process defined by the brain's modelling of potential causes for sensory inputs. An interpretative prediction is generated on the basis of incoming signals; this prediction is refined in the light of experience, expectation and fresh information; and this looped process of feedback and refinement is folded into perception's ever-flowing stream. The result is not so much an unvarnished account of the external world as a series of mental models connected to external causes in *useful* ways. As the cognitive scientist Donald Hoffman puts it in the preface to his 2019 book *The Case Against Reality*:

> [Evolution] has endowed us with senses that hide the truth and display the simple icons we need to survive long enough to raise offspring . . . Your senses have evolved to give you what you need. You may want truth, but you don't need truth. Perceiving truth would drive our species extinct.[4]

In evolutionary terms, not all perceptions are created equal – while utility is a more precious quality than completeness. A version of vision that kept me constantly, identically aware of every inch of the world would be at once overwhelming and incoherent. What I need is an avowedly selective, partial way of perceiving that enables me to identify potential threats; to spot sources of food and shelter; to accurately predict the behaviours of kin, prey and rivals; to relegate extraneous information to the background; and so on. An actionable apprehension of these things is vital. But at no point does this entail raw sensory inputs being displayed to some miniature decision-making 'me' inside my head.

It's worth emphasizing just how hard it can be to embrace the implications of these facts – and how deeply embedded the idea of perception as a linear process is in our everyday thinking. Consider what happens when I spot my pet cat in the garden. A typical explanation of the resulting sequence of neurological events might go something like this. First, light bounces off the cat and enters my eye, casting its image onto my retina. Next, information about this image is transmitted along the optic nerve into my brain. This image is then unscrambled and 'looked up' by various remembering and controlling modules until, finally, the thought 'hey, that's my cat!' slips into my stream of consciousness.

All of this makes intuitive sense. Yet it's not only incorrect in almost every detail, but also – thanks to the problem of an infinite regress – doesn't actually constitute an explanation at all. If a tiny 'me' is metaphorically seated inside my skull, pondering representations of the world, does that mean there's an even tinier 'me' metaphorically sitting inside its metaphorical skull, and so on? Perception cannot be explained by positing any number of inner 'control' modules – because we would then need to go inside these modules to do any actual explaining, only to find ourselves back where we started.

What words and ideas should we deploy to describe what's actually going on? Instead of imagining a 'Cartesian Theatre'[5] within which a miniature 'us' sits watching a screen, Seth suggests that the non-linear nature of our experiences can be captured by the concept of 'controlled hallucination', a phrase he notes first hearing from the British psychologist Chris Frith.[6] Here is Seth's summary of what this means:

The essential idea is that the brain is a prediction machine, so that what we see, hear, and feel is nothing more than the brain's best guess of the causes of its sensory inputs. Following this idea all the way through, we will see that the contents of consciousness are a kind of waking dream – a controlled hallucination – that is both more than and less than whatever the real world really is . . . it is natural to think of perception as a process of bottom-up feature detection – a 'reading' of the world around us. But what we actually perceive is a top-down, inside-out neuronal fantasy that is reined in by reality, not a transparent window onto whatever that reality may be.[7]

Crucially, by replacing the idea of perception as a 'reading' of the world with the claim that it's an actively constructed 'neuronal fantasy', Seth swaps the standard dichotomy of hallucination/reality for something more like a spectrum: a gamut of perceptual reliability running from uncontrolled to controlled.

If, under the influence of a psychoactive drug, I imagine that a picture of my pet cat has come to life and is talking to me, I am experiencing an uncontrolled perception – because the drug has disrupted the dampening effect of incoming sensory feedback on my brain's interpretative inferences. But even in my soberest and most literal-minded moments, my perceptions are still a fabrication comprising best-guesses, preferences and projections. The degree to which these are 'controlled' by features of the external world is simply very high. As Seth puts it:

In 'normal' perception, what we perceive is tied to – controlled by – causes in the world, whereas in the case of [uncontrolled] hallucination our perceptions have, to some extent, lost their grip on these causes . . . You could even say that we're all hallucinating all the time. It's just that when we agree about our hallucinations, that's what we call reality.[8]

And all this is as it should be. How else could I, or anyone else, have evolved to think and act rather than endlessly prevaricate? Just like every healthy human being, I jump at sudden noises; I see human-like forms in the branches of a tree; I sketch fears and phantoms onto the canvas of sleep. We all move through perceptual worlds charged with meanings and purposes, because anchoring our minds to the world is a pragmatic, probabilistic project within which the prospect of perfect clarity is at once impossible and incoherent.

This chapter's subtitle, *The delusion of literal-mindedness*, addresses the temptation to pay attention only to the surface of these things: to adopt an appealing but inaccurate 'mini-me' view of the mind, according to which some ineffable inner decider lurks inside my head weighing up what 'I' should do next. In the context of neuroscience, it's a model whose intuitive appeal cannot be squared with scientific knowledge. In the context of technology, it's a form of oversimplification that ignores the most remarkable features of our mental lives: our uneasy mixture of self-division, unreliability and iterative potential; our capacity to become less deceived about both ourselves and the world *because of*, not despite, the protean nature of our apprehensions.

Putting perception to a meaningful test

Think back to the Müller-Lyer diagram with which this chapter began. We call it an illusion because it creates a perception that's at odds with empirical reality. Two identical lines appear two different lengths because we can't switch off the active, partial processes through which perceptions are generated. Yet – as the very phrase 'empirical reality' suggests – our *assessment of this perception* is eminently amenable to insight, not to mention research, exploitation and augmentation. And it's this informed negotiation with perception that above all characterizes the many-layered wonder of human consciousness.

Nowhere is this more evident than in the integration of technology into perception. I look out of the window at my garden and I see trees swaying in the breeze. I take a video with my phone and watch it. Once again, I see swaying branches. Yet what I'm watching isn't, technically, a moving image: it's a series of static images which, when thirty or more of them are shown each second, creates a compelling illusion of movement. The phenomena underpinning this, sometimes bracketed under the heading 'the persistence of vision', remain imperfectly understood – but that hasn't stopped us from exploiting them to create entire worlds of mediated experience.[9]

If I slow down the rate of video playback, the illusion of motion is gradually replaced by an awareness of the gap between each frame. Depending on what you're filming, things start to stutter at around sixteen frames per second. If I put my phone into slow-motion mode, I can film at 120 or even 300 frames per second. Playing back these films creates the illusion that time is moving four or ten times more slowly than usual. Yet this illusion is no more or less real, in its raw mechanics, than any other recording. What's 'actually' going on when I watch any video is that a human being is observing a rapidly shifting series of bright dense pixels on a flat screen.

Or is it? As Seth's account of perception makes clear, the most significant distinction when it comes to consciousness isn't between illusion and reality at all. After all, it's not like we ever experience (or could hope to comprehend) reality in its naked form. Rather, the relationship that matters most is one that's particularly significant in the context of this book: between *illusion* and *delusion*. It's about the gulf between how things appear and what we believe their truth to be.

In his 2011 book *The Beginning of Infinity*, the physicist David Deutsch describes how, as a graduate student, he pored over photographic negatives from the Palomar Sky Survey, which showed stars and galaxies as dark shapes on silvered glass plates, clustered so minutely that microscopes were required to examine them. At one point, a tiny defect in the image led him briefly to imagine he had discovered a new galaxy. A speck of dust on a telescope's lens momentarily stood in for billions of stars.

What, Deutsch asks himself in the light of this confusion, was actually happening when he peered through his microscope's lens at a photographic image of a telescope's capture of ancient light. Was he looking at distant galaxies, or at specks of silver on glass? Thinking back, Deutsch has no doubt. Both things were simultaneously true. 'I was indeed looking at galaxies,' he writes:

> Observing a galaxy via specks of silver is no different . . . from observing a garden via images on a retina. In all cases, to say that we have genuinely observed any given thing is to say that we have accurately attributed our evidence (ultimately always evidence inside our own brains) to that thing.[10]

Deutsch's eyes may have been focused on earthly artefacts, but his mind was focused on wonders half a universe away. And, being equipped with an adequate method for sifting delusions from illusions, he could grasp these as surely as anything else in the realm of perception. Indeed, he suggests, scientists' use of technology to glimpse staggeringly distant objects is simply an extreme case of something that's true of every single experience. Because we cannot help reading the world in more than merely literal terms, what counts is not so much what we're attending to – near or far, abstract or concrete – as how we integrate it into our understanding.

> It may seem strange that scientific instruments bring us closer to reality when in purely physical terms they only ever separate us further from it. But we observe nothing directly anyway. All observation is theory-laden. Likewise, whenever we make an error, it is an error in the explanation of something. That is why appearances can be deceptive, and it is also why we, and our instruments, can correct for that deceptiveness.[11]

In the end, Deutsch's illusory galaxy tells us the same thing as every conscious moment: that perception is at once predicated upon predictable illusions and able constantly to recontextualize these illusions in the light of knowledge. Similarly, technology doesn't just enhance our intellect or powers of observation: it's also our partner in the dance of theories and preferences through which we conjure a coherent world.

The dynamic, divided self

In one sense, being able to see distant galaxies numbers among humanity's most staggering achievements: the culmination of millennia of scientific, technological and intellectual progress. In another sense, however, it's a beautifully straightforward proposition. Given the right instruments and approach, millions of galaxies can now be identified and enumerated with a high degree of consensus. In a limited but remarkably reliable way, we can extend the compass of our minds across billions of light years. By contrast, it's exquisitely difficult to observe something that couldn't be closer to hand: ourselves.

The philosopher David Hume famously spelled out the slippery nature of the self in his 1739 *Treatise of Human Nature*. Although they're almost three hundred years old, his observations still deserve close reading thanks to their alertness to self-perception's central paradox: that we are never *not* perceiving, and thus that all attempts to catch an inner 'us' doing this perceiving are thwarted by the doubling of subject and object this implies.

> For my part, when I enter most intimately into what I call myself, I always stumble on some particular perception or other, of heat or cold, light or shade, love or hatred, pain or pleasure. I never can catch myself at any time without a perception, and never can observe any thing but the perception . . . I may venture to affirm of the rest of mankind, that they are nothing but a bundle or collection of different perceptions, which succeed each other with an inconceivable rapidity, and are in a perpetual flux and movement.[12]

It's after this passage in Hume that the so-called 'bundle' theory of the self is named. The theory anticipates Seth's observations about perception – and emphasizes that they apply as much to the alleged 'self' doing the perceiving as to what's being perceived.

Much as perceptions appear to grant some inner 'me' access to the external world, but turn out upon closer inspection to constitute an unending series of feedback loops, so the selfhood these processes conjure is *itself* composed of the same stuff. Like a video playing on my phone, it's a flickering series of frames that usefully cohere into unity not because there is any fixed, observing 'me' inside my head, but because the impression that one exists is a marvellous technique for navigating reality. Along with everything else in the realm of controlled hallucination, 'I' am a constantly self-correcting series of predictions and inferences – a process interlaced with countless other processes. As the philosopher Julian Baggini puts it in his 2011 book *The Ego Trick*:

There is no place in the brain where it all comes together and there is no immaterial soul which is the seat of consciousness. The unity we experience, which allows us legitimately to talk of 'I', is a result of the Ego Trick – the remarkable way in which a complicated bundle of mental events, made possible by the brain, creates a singular self, without there being a singular thing underlying it.[13]

At this point, it's worth pausing to spell out the analogy between our individual efforts to grasp the world and the collective project of reasoned comprehension. In each case, the most important point is that neither discrete, linear processes nor final answers are adequate analogies to what's actually going on. No verdict can ever be exhaustive or free from presumption; no theory tenable in the absence of testing and iteration. The world is what we make of it in the most literal of senses: an illusion within which layer upon layer of truth can be discerned.

Indeed, even words like 'introspection' imply metaphors that break down when it comes to our mental processes – for I am neither *looking* nor *seeing* when I try to find words for the flow of my thoughts, nor can these thoughts ever be *visible* or *obscure* to me in anything like the manner of sense perceptions. The more we try to describe our mental lives in such terms (*'An image of the cat came in through my eyes, my brain looked at it, then I decided to give it some food'*), the more the very words we're using obfuscate the mechanisms at work – not to mention the true nature of our relationships with and through the systems surrounding us.[14]

What should we take from this? Among other things, that which seems simple and self-evident often turns out to have little to do with reality – and everything to do with the mechanisms that help us navigate it. As the psychiatrist and author Ian McGilchrist has emphasized in his work on the relationships between our minds and the world, we can unearth and explore all manner of facts *about* our shared existence. But we confuse these with actuality itself at our peril. Our mental models, tidy though they may be, are quite different to the vastness of what's out there.

> Complexity and simplicity are relative terms. However, complexity is surely, we imagine, a more unusual state of affairs arising out of the agglomeration of more simple elements – isn't it? I believe that this is a mistake . . . Rather, complexity is the norm, and simplicity represents a special case of complexity, achieved by cleaving off and disregarding almost all of the vast reality that surrounds whatever it is we are for the moment modelling as simple (simplicity is a feature of our model, not of the reality that is modelled).[15]

McGilchrist's argument, here, echoes some of the most fundamental insights of the sacred Hindu texts, the *Upanishads*: that both the world and the self are in a ceaseless state of flux; that the reality we are a small part of embodies limitless, incomprehensible complexity; and that it's certainty and fixity, not change and complexity, that are the wishful illusions.[16]

This may seem, to say the least, an abstract insight. Yet it is directly related to one of the twenty-first century's most urgent technological issues: the nature and the ethics of *attention*. In particular, it can help us analyse the ways in which we often seek – and are sought out by – forms of simplicity that 'cleave off' much of what makes minds and the world so complex; not to mention how we might restore some vivifying complexity to both our thinking and the systems with which it's interlaced.

What I pay attention to becomes my life

How many other things could you be doing right now? How many messages, notifications, updates and social media requests will be waiting for you once you put this book down? The selectivity and elasticity of human attention is a constant feature not only of perception but also of cognition, belief and identity. And this in turn suggests why attention is a prized commodity in the realm of media and technology: because it so powerfully shapes our beliefs, inclinations and interests. Those who command our attention can aspire both to predict our actions and to make their predictions come true. As the writer Charlie Warzel put it in the *New York Times* in February 2021, in an article exploring the author Michael Goldhaber's work on attention:

> One of the most finite resources in the world is human attention . . . Every single action we take – calling our grandparents, cleaning up the kitchen or, today, scrolling through our phones – is a transaction. We are taking what precious little attention we have and diverting it toward something. This is a zero-sum proposition . . . When you pay attention to one thing, you ignore something else.[17]

For all the urgency of debates about attention's measurement and enticement, however, even the framing of this discussion in terms of 'resources' and 'zero-sum' propositions can stop us from attending to its most significant elements. Why? Because human attention is most precious not because it embodies new frontiers for data and profit, but because it's via the quality and nature of our attending that different worldviews are weighed in the balance.

In effect, the manipulation of attention is about someone else trying to define the terms of your relationships with the world: that which counts and that which is passed over. By 'attention' we mean nothing less than the place where empathy and care are offered or withheld; where introspection and imagination work their transformations. What I pay attention to becomes my life – and this means that there is a fundamentally *moral* element both to the attention I pay things and to the kinds of attention that they seek from me.[18] As McGilchrist put it in a 2011 article for *The Lancet*:

> Attention is reciprocally related to what exists: it's not just that we attend differently depending on what we find, but that what we find depends on the kind of attention we pay.[19]

When combined with the insights of the extended mind hypothesis, it's clear that the kind of attentional engineering embedded in different technologies poses significant questions not only about individual thriving, but also about the self-knowledge and capacities for self-correction of society as a whole.

What assumptions and predictions are our minds being pushed towards by the systems they're enmeshed in: by their interfaces, abstractions and automations? What might it mean for our divided, extended minds to comprehend the challenges of our age more richly; to reach out through concepts and tools that enhance rather than constrict our autonomy, compassion and insight? Such questions sketch a task that will define much of our future: how to create cognitive artefacts capable of delivering us from delusions rather than snaring us within them. Literal-mindedness – a way of thinking that denudes our existence of depth and complexity – is the last thing we can afford.

CHAPTER 6

How technologies invent themselves:
The delusion of comprehension

Thus far, this book has looked at humanity's co-evolution with technology in foundational terms. It has suggested that technology predated the very existence of *Homo sapiens*; that its existence amongst hominins was bound up with the faculties of imagination, communication and collaboration; and that the transformation of tools into technologies paralleled our ancestors' emergent ability to think not only about how things are, but about how they *might* or *could* be: about what it might mean to remake both the self and the world.

It has also suggested that all this continues to be interwoven with humanity's unique plasticity of mind and vulnerable, extended childhood. Our children have the potential to adapt, learn and grow across years of dependence, sustained by networks of loving cooperation; and this has opened up opportunities for survival and thriving far outstripping the rest of our primate order. Heated by domesticated flames, linked by language, we are the sole survivors of our lineage – and lucky to have made it thus far. More inquisitive and vulnerable than our vanished kin, more restless in our exploration and remoulding of our environment, we are embarked upon an exponential path towards something unprecedented: towards a planetary future whose hopes and hazards rest in our hands.

As a species, our deep interest in the dynamics of cooperation plays a central part in all this, both as an enabler of selfless norms and of brutally divisive forms of envy and acquisitiveness. Unlike all other species, we care less about sufficiency than comparison. There is no point at which, collectively, our imaginations and ambitions say, 'Enough!' But there is something that binds us to reality: a deep interest in making sense of our world that's intimately linked to our survival thus far and the prospect of our endurance.

Perhaps the most crucial distinction here is between *illusion* and *delusion*. Our minds and perceptions are theory-laden. We do not and cannot access the external world directly or objectively. Rather, we are able to theorize about its workings in ways that are at once extraordinarily sophisticated and moulded by multiple modes of attention. Our stone age brains are all too open to error and manipulation: to ideologies that pit tribes against truth, instinct against apprehension. Yet we also have the capacity to extend our mental lives far beyond our bodies; to reach across interstellar distances and decipher our universe's most fundamental features.

Every insight and opportunity has its shadow. We cannot afford to treat technology as a form of destiny, mastery as our species' entitlement, or evolution as a blood-soaked battlefield. And before we can look to the future, there is one last self-deception that needs to be guarded against. We need to push back against the *delusion of comprehension* – and to grasp, alongside the scale of our achievements, the belatedness of our understandings.

How the sea designs a boat

In his 2017 book *From Bacteria to Bach and Back*, the philosopher Daniel Dennett presents a potted history of our minds' emergence from life's most basic building blocks. Here is his summary of the last few million years.

> Human culture started out profoundly Darwinian, with uncomprehending competences yielding various valuable structures in roughly the way termites build their castles, and then gradually de-Darwinized, becoming ever more comprehending, ever more capable of top-down organization, ever more efficient in its ways of searching Design Space. In short, as human culture evolved, it fed on the fruits of its own evolution, increasing its design powers by utilizing information in ever more powerful ways.[1]

Dennett's analysis captures both the revolutionary nature of the mental developments explored in the second chapter and the recursive, accelerating impacts of technology itself. Thanks to the feedback loops engendered by discoveries like writing, printing and computation, modern humans can explore the 'design space' of ideas and possibilities in the light of unprecedented knowledge. Yet, as Dennett is also at pains to point out, even the most recent incarnations of this activity haven't so much freed us from worldly constraint as shifted the terms of an immemorial negotiation.

Consider the case of a simple wooden fishing boat: one carved by a villager and used to cross the waters of a great bay. The villager based the design on boats sailed by older villagers, as they did theirs. Nobody in this fishing community is what you might call a designer or an engineer, and the boats they make are informed by a reverence for tradition. Yet, over time, tiny design improvements accumulate: raised bows and weighting beneath the waterline, improving seaworthiness; protective treatments that help keep the hull watertight and free from decay. How does this happen?

One answer to this question can be found beneath the waters of the bay, where the silted remnants of sunken boats lie. These are the boats that didn't bring their makers safely home. Each new boat is made in imitation of an older boat – and, once lost, no boat can be imitated. Thus, over time, the advantages or weaknesses of even minor design variations exert an evolutionary force. As Dennett puts it, in a passage glossing observations first made by the French philosopher Émile-Auguste Chartier:

> 'One could then say, with complete rigor, that it is the sea herself who fashions the boats, choosing those which function and destroying the others.' The boat builders may have theories, good or bad, about what would be an improvement in the craft their fathers and grandfathers taught them to build. When they incorporate these revisions in their own creations, they will sometimes be right and sometimes wrong. Other innovations may simply be copying errors, like genetic mutations, unlikely to be serendipitous improvements, but once in a blue moon advancing the local art of boatbuilding.[2]

To make the picture more complex, survival is not the only thing that counts – because a technology's fitness resides not only in the world, but also in the attitudes and experiences of its creators. Anything that does no harm while also pleasing those who make boats is likely to gain a foothold over time: decorations, names, creature comforts. Harmless adornments may be taken for benefits, or causal relationships inferred where none exist: the naming ceremony is said to safeguard the boat, the decorations to ward against evil.

Are undertakings like this useless in evolutionary terms? Not necessarily. If they confer psychological benefits, they may genuinely contribute to the survival and thriving of the boats' creators. Naming and launching ceremonies may bind a community together, or promote care and dignity in the business of boat-building. Songs may memorably encode advice about where and when to fish, what to beware of at sea, or when to come home. In the long term, the refinement of meanings and purposes – of symbolic and spiritual resonance – may count most of all, outliving the age and the culture that birthed them.

The evolutionary dynamics at work beneath the boat-builders' notice resemble those shaping every one of the Earth's species. To borrow Dennett's phrase, evolution bristles with cases of 'competence without comprehension' – and features of this remain the case even when purely Darwinian processes have been left behind.[3] Much as biological organisms can develop stunningly complex behaviours without 'knowing' anything about their ultimate objectives, it's possible for a boat-building society to become adept at constructing fast, seaworthy craft without a formal understanding of fluid dynamics, hull design or weather systems. Indeed, most technologies tend over time to 'improve' in countless unintended ways thanks to evolution's underlying diktat: that, given variety, selection pressure and some form of heritability, adaptations which support survival and reproduction will emerge.

What does this mean for our understanding of technology? For a start, it suggests that resisting determinism mustn't mean either ascribing perfect insight to our ancestors or aspiring towards it ourselves. Did some ancient hominin look at a sharp rock, once upon a time, struck by the revelation that it might come in handy? Almost certainly not. The development of a tool or innovation may seem to imply the identification of a problem awaiting a solution: a lock waiting for a key. But the workings of time and chance look little like this. Countless rocks opportunistically handled in countless ways will, eventually, yield an outcome beneficial enough to be worth repeating and imitating. And the degree of understanding such developments require is more minimal than that implied by tales of 'eureka!' moments.

To insist that progress is synonymous with solution-seeking, in other words, is to substitute retrospective tales of cause and effect for something more equivocal. As the example of the fishing boats emphasizes, uncomprehending evolutionary forces – the only ones present in the wholly 'Darwinian' context of our earliest evolution – continue to be felt *even* in situations where a thoughtful, sophisticated culture exists. And it is these forces that provide the ultimate test against which our efforts will be judged.

At stake here is a subtle point that's worth teasing out. We have evolved far beyond our uncomprehending origins; indeed, we have an astonishing capacity to work and strive within the realm of theory, extending and recombining present possibilities constrained only by imagination and reason.[4] At the same time, however, we are inclined to treat every discovery as a heroic narrative of individual foresight and problem-solving – while ignoring both the communal, technologized character of 'design space' and the prospect that our most cherished schemes may prove less than prophetic.

It's comforting to believe that world-changing events can be traced to personal acts of will and revelation, and that all the world can be explained in line with human cognitive preferences. But this rationalization of events into anthropocentric narratives is a gift evolved for utility rather than accuracy; and among its unfortunate consequences is a tendency to fabricate human-scale reasons even when such things don't exist. From animal behaviour to weather systems, from mindless automata to the inscrutability of our own unconscious urges, we project a transparent and fully explicable idealization of our own mental life onto the world. And by doing so we risk ignoring the best lessons that scepticism, doubt and uncertainty might teach. As Dennett puts it:

> The normativity of reason-giving imposes itself even when we are at a loss for an answer. There is an obligation to have reasons that you can give for your behaviour . . . we lead our daily lives bathed

in the presumption of understanding, in the strangers we encounter as well as our family and friends. We expect people to expect us to have reasons, reasons we can express, for whatever it is we are trying to do, and we effortlessly make up reasons – often without realizing it – when a query catches us without a pre-thought answer.[5]

The delusion of comprehension, in other words, is closely connected to that of literal-mindedness, with its insistence that intuitive, human-centric stories are the only accounts of actuality we need.

Perhaps the greatest challenge of such a delusion is that it flatters our gifts while effacing our limitations: that it encourages us to narrate our lives according to a satisfying succession of rationalizations while refusing to grapple with the *actual* complexities of consequence and chance. Can we do better? Absolutely. But only if we embrace forms of attentiveness and open-mindedness that are alive to the unexpected and the uncertain: to the lessons lurking within time's losses.

The significance of survivorship

Let's look one last time at the case of the fishing boats. Hidden beneath the tides lie the remnants of those that sank. What role should these lost boats play in the thinking of those crafting new ones: the stories they tell, the skills they pass on? In evolutionary terms, they are this culture's hidden history. Like our extinct ancient ancestors, they map what *might* have been. Their failure memorializes the most important ways in which fate brought the present into being, and this means that any lessons drawn from the past will be incomplete if they are forgotten.

Now imagine a huge storm striking the fishing village. It's a tragedy of existential proportions for this community: just two boats out of dozens survive. What was so special about these two, their builders ask? Was it the survivors' high bows; the courage of their captains; their hulls' sacred markings? If these are the surviving boats' most striking features, they're likely to be praised and imitated. Yet beneath the water sleeps a more mundane truth. Many sunken boats also had all these features; but only the survivors held heavy catches in their hulls. Their stabilizing effect was most likely what kept some boats afloat. But history, if told as an annal of heroism and endurance, has little time for lessons like this.

The general principle this illustrates is known as *survivorship bias* – and one of the most famous examples of how it can be redressed is the statistician Abraham Wald's analysis of Allied aeroplane losses during the Second World War. Born in the Austro-Hungarian Empire in 1902, Wald studied at the University of Vienna before becoming director of the Austrian Institute for Economic Research. As the political situation in Europe began to worsen, he emigrated to America, where he ended up working at the classified Statistical Research Group programme in Manhattan. Among other projects, Wald was asked to analyse US military data about the distribution of bullet holes on aircraft returning from combat in Europe. How, the military wanted to know, should armour most efficiently be distributed across a plane in order to protect it while keeping its weight to a minimum?

Wald's investigation of the data led to a recommendation that ran directly counter to previous thinking. The armour should be reinforced wherever there were fewest bullet holes, and reduced wherever there were most. Why? Because counting the bullet holes on returning planes told less than half the story. The most noteworthy planes were not the survivors, but rather those that *never* returned: those whose wreckages lay scattered across Europe. As the author and mathematician Jordan Ellenberg notes in his 2014 book *How Not To Be Wrong*, it's a story that suggests some vivid analogies.

> If you go to the recovery room at the hospital, you'll see a lot more people with bullet holes in their legs than people with bullet holes in their chests. But that's not because people don't get shot in the chest; it's because the people who get shot in the chest don't recover.[6]

This is the significance of survivorship writ large. Human reasons and anticipations allow for remarkable kinds of planning. But the 'reasons' that matter most of all are those provided by time and events – and the most precious lessons are those that an alertness to hidden histories can glean from these.

Achieving this alertness is often counter-intuitive. Consider the following question: how meaningful is it to itemize the attributes of a successful company based upon the exploits of Apple or Google, or a successful entrepreneur based upon the habits of the world's richest people? Although such analyses make for lively reading, they are largely worthless when it comes to generating insights. Why? Because they ignore the legions of unsuccessful companies and individuals that did some or all of the same things. The sheer notability of the survivors, coupled to the relative invisibility of the failures, means that such tales can prove irresistible. Yet they're little more useful than trying to explain how someone won the lottery based on what they had for breakfast.

Evolutionary processes present us with a constant version of this problem. As the author and risk expert Nassim Nicholas Taleb notes in his 2001 book *Fooled by Randomness*, time is itself a deadly game. If enough people play high-stakes Russian roulette every year, after a few decades a few extremely wealthy individuals will be left standing. This doesn't make playing annual games of Russian roulette a sensible strategy. But it may well seem like one to an observer.

> If a twenty-five-year-old played Russian roulette, say, once a year, there would be a very slim possibility of his surviving until his fiftieth birthday – but, if there are enough players, say thousands of twenty-five-year-old players, we can expect to see a handful of (extremely rich) survivors (and a very large cemetery).

Moreover, Taleb continues:

> Reality is far more vicious than Russian roulette. First, it delivers the fatal bullet rather infrequently, like a revolver that would have hundreds, even thousands, of chambers instead of six . . . Second, unlike a well-defined, precise game like Russian roulette, where the risks are visible to anyone capable of multiplying and dividing by six, one does not observe the barrel of reality.[7]

Managing to look down the 'barrel of reality', in Taleb's memorable phrase, is a profoundly different prospect to parsing pre-packaged narratives, which can conjure a comforting illusion of understanding amid even the greatest uncertainties. In particular, doing so means confronting the countless failures of survival and adaptation that shadow every success: the dead ends from which evolution's illusion of purpose emerges; the catastrophic risks from which there can be no coming back.

Let's return, for a moment, to Abraham Wald's story. You may have heard it before. Indeed, you may well have groaned with familiarity when encountering it. It has become a staple of business school seminars, its lessons a PowerPoint deck, diagrams derived from it textbook clichés. Hearing it for the first time tends to provoke a satisfying 'eureka!' moment, and for good reason. It's a parable of a heroic individual achieving a breakthrough against the odds: a narrative meme rendering an abstract insight tractable. Yet, as the author Tim Harford points out in his 2020 book *How to Make the World Add Up*, this is because Wald's analysis is *itself* subject to a powerful form of survivorship: our tendency to turn the past into memorable stories, in the process stripping away layers of context and complexity. As Harford puts it:

> What survives is the tale about a mathematician's flash of insight, with some vivid details added. What originally existed and what survives will rarely be the same thing.[8]

Fortunately, 'what originally existed' is at least partly recoverable in the form of Wald's original research paper, a document that belies anecdotal neatness on every page. Across hundreds of lines of statistical analysis, he pieces together a counter-intuitive point from first principles.[9] There are no diagrams, no folksy generalizations and no hints of self-aggrandizement. At a time of global conflict, amid profound uncertainty, Wald sought to approach questions of life and death by systematically accounting for failures and accidents: not so much a lone hero as an alert mind in the right place at the right time, equipped with relevant and accurate information, pooling the insights of multiple fields.[10]

Much as Gutenberg had brought together paper, ink, metallurgy, wine-presses and the Roman alphabet centuries before, so Wald brought to fruition a project predicated upon certain skills, opportunities and information – as well as sufficient attentiveness to tease out their patterns. Indeed, the capacity simply to *notice* what's going on is among the most precious of all attributes when it comes to analysing technology: a foundational skill that often comes more easily to children than their parents.[11]

What can we learn from Wald's example? Most obviously, people tend to be preoccupied by lucky survivors – and survivorship needs to be considered in almost every real-world situation. Less obviously, thinking sufficiently rigorously about fate and failure is precious and rare. It requires time, focus and cognitive reinforcements: a combination of relevant information, enabling framings and rigorous purposes that embody the labour and permission of many systems and minds.

How to learn from accidental revelations

In his 2005 book *The Original Accident*, the French philosopher Paul Virilio offers a provocatively paradoxical account of how innovation can simultaneously embody new opportunities and create new kinds of crisis.

> To invent the sailing ship or steamer is to invent the shipwreck. To invent the train is to invent the rail accident of derailment. To invent the automobile is to produce the pile-up on the highway . . . It follows that fighting against the damage done by Progress above all means uncovering the hidden truth of our successes in this accidental revelation . . .[12]

This notion of the 'accidental revelation' echoes the hidden histories explored in the previous section. Humanity surveys evolution's obstacle course – to return to the metaphor of the second chapter – from a position of unique insight. Yet the complexities and cascading consequences of our creations have never been more unpredictable, nor certain accidents more likely to prove catastrophic. Hence Virilio's evocation of 'the damage done by Progress', which captures something of novelty's carelessness and fragility. It gestures beyond the battle cry of tech disruption – 'move fast and break things!'[13] – towards the belated knowledge that breakage can bring.

Another way of putting Virilio's point is that while the insights brought by accidents may be impossible to predict, this only makes it more important to seek out such revelations in as timely and honest a manner as possible; and to ensure that the 'design space' within which we operate is informed by an ethically and statistically literate approach to risk. This in turn means acknowledging *both* that the unintended side effects of technological modernity define our most urgent challenges *and* that labelling them 'unintended side effects' is part of the thinking we need to overturn. As John D. Sterman, director of the MIT System Dynamics Group, put it in a prescient 2006 article exploring the significance of systems thinking in the context of contemporary complexities:

> There are no side effects – just effects. Those we expected or that prove beneficial we call the main effects and claim credit. Those that undercut our policies and cause harm we claim to be side effects, hoping to excuse the failure of our intervention. 'Side effects' are not a feature of reality but a sign that the boundaries of our mental models are too narrow, our time horizons too short.[14]

Sterman's article was first published in the *American Journal of Public Health*, and its conclusions have a bleak relevance in a world still reeling from the Covid-19 pandemic, not to mention its associated crises of trust, comprehension and control. The nested challenges of globalized modernity defy our understanding, Sterman suggests, not only because grasping them is inherently difficult, but also because the very notion of achieving a definitive, solution-orientated understanding doesn't make sense in an age of systems too complex and mutually embedded to predict.

> When humans evolved, the challenge was survival in a world we could barely influence. Today, the hurricane and earthquake do not pose the greatest danger. It is the unanticipated effects of our own actions, effects created by our inability to understand the complex systems we have created and in which we are embedded. Creating a healthy, sustainable future requires a fundamental shift in the way we generate, learn from, and act on evidence about the delayed and distal effects of our technologies, policies, and institutions.[15]

What should the 'fundamental shift' Sterman calls for entail? So far as technology is concerned, it means attending to the human-made world as something entwined with our lives and minds rather than optional or additional to them. And this in turn means confronting the flip side of technology's enhancement and interconnection of these minds: those forms of cultural and technological evolution that serve purposes inimical to our thriving; the junk and the spam; the conspiracies and manipulations that can so vigorously colonize the mechanics of human sense-making.

In particular, to return to Plato's enduring insight into the cost of deceptions, we need to ask how and why information technologies may become aligned with the denial of reality: with manipulative and authoritarian forms of delusion. While Plato was most deeply concerned with untruth, however, perhaps the deepest structural challenges we face today entail an *indifference* to truth. That is, they entail forms of persuasion and coercion committed not so much to lies as to a world-view where truth-seeking is irrelevant, and where the 'facts' of any matter are whatever those wielding sufficient influence declare them to be. We are back in the realm of attention and oversimplification – and the ways in which information systems shape not only *what* we think about but also *how* and to *what purpose*.

Overcoming bullshit

In his 2005 book *On Bullshit* – which first appeared in the form of an essay in the Fall 1986 issue of *Raritan* magazine – the late philosopher Harry Frankfurt makes a strikingly prescient argument. 'The essence of bullshit', he writes, 'is not that it is false but that it is phony.' He continues:

> It is impossible for someone to lie unless he thinks he knows the truth. Producing bullshit requires no such conviction. A person who lies is thereby responding to the truth, and he is to that extent respectful of it. When an honest man speaks, he says only what he believes to be true; and for the liar, it is correspondingly indispensable that he considers his statements to be false. For the bullshitter, however, all these bets are off: he is neither on the side of the true nor on the side of the false . . . He does not care whether the things he says describe reality correctly. He just picks them out, or makes them up, to suit his purpose.[16]

Both a liar and an honest person are interested in the truth: they're playing on opposite sides in the same game. A bullshitter, however, has no such constraint. So far as they are concerned, an assertion's only purpose is its effect on others. All that matters is the degree to which it does, or doesn't, further their personal ends.

Imagine a politician who claims to have witnessed something that did not, in fact, happen: thousands of people in areas of New Jersey with a heavy Arab population cheering as the World Trade Center came down on September 11th 2001. Now imagine an interviewer confronting them with evidence that their claim is untrue – as George Stephanopoulous did to Donald Trump in November 2015, during the course of ABC's *This Week*.[17]

The rules of the game the interviewer is playing dictate that a politician who has been caught in a lie should offer an apology, an excuse, or some compelling new evidence of their own. Instead, this particular politician invoked another strategy: contempt for the rules of the whole reality-based game. 'It did happen. I saw it,' Trump replied. 'It was on television. I saw it.' There was no evidence to support his claim, yet he continued: 'It was well covered at the time, George. Now, I know they don't like to talk about it, but it was well covered at the time.' It wasn't, but this didn't matter. Truth may be forceful against lies, but it bothers bullshit about as much as paper darts do a tank.

Bullshit, propaganda and naked untruth have immemorially been the stuff of politics. The twenty-first century, however, is experiencing the consequences of affective impact becoming not only the supreme test of a story but also part of the fabric of information technologies themselves. We inhabit, much of the time, information environments that are at once world-spanning and structurally indifferent to truth. They are far from indifferent, however, to one of the most important evolutionary pressures for any information technology: the emotional state of its users.

The ecological overtones of the phrase 'information environment' aren't accidental. News and narratives – fake and otherwise – exist in constant, ferocious competition for belief and engagement. In our age of information suffusion, their supply is plentiful while the attention upon which they thrive is scarce. The result: an evolutionary hothouse within which the fittest fabrications leap between clicking fingers and glancing eyeballs while the weak indifferently wither.

Once again, it's a mistake to overestimate the degree of intention and control at play. For all that tech companies, hackers and scammers are portrayed as master manipulators, it doesn't take superhuman skill to send a lie skipping around the world: simply the shameless repetition of whatever people are already inclined to believe, using whatever means have already proved themselves effective. Similarly, the technologies that mediate and instantiate this aren't just implementations of their creators' intentions. They're also a series of interlocking, evolving systems whose comprehension-free competences have a life of their own – or something remarkably like it.

Fitness, here, is a question of adaptation to a miscellany of incentives: those engineered by the companies who make and profit from information systems; those baked into human brains by millennia of evolution; those born from coincidence, serendipity and survival. The result? A festival of bullshit – of attention won at all costs – within which disinformation is not so much the sinister cause of present ills as the price of admission for an informational era. As Dennett puts it towards the end of *From Bacteria to Bach*, in a passage drawing on Richard Dawkins's definition of a *meme* as any 'unit of cultural transmission':

> Many memes, maybe most memes, are mutualists, fitness-enhancing prosthetic enhancements of our existing adaptations . . . But once the infrastructure for culture has been designed and installed [i.e. evolved in human minds] . . . the possibility of parasitical memes exploiting that infrastructure is more or less guaranteed.[18]

Such memes are viral in their spread, a metaphor so ubiquitous that we no longer notice it's metaphorical. Indeed, the analogy with biological viruses is disturbingly apposite. In each case, the entity in question has no independent life of its own. It is not only parasitic but also non-functional outside of its host: a stripped-down replicator able to evolve and reproduce only when it becomes part of another living thing.

As the Covid-19 pandemic made clear, the most pernicious of biological and informational parasites thrive upon inequality and vulnerability. Infections pour through the cracks in society, disproportionately affecting the marginalized, the dispossessed, the desperate; those living precariously; those isolated or alienated from structures of mutual care and support. The confluence between physical and informational susceptibilities creates fresh opportunities for infection: those convinced that mainstream beliefs are conspiracies; that the state is lying to them; that personal freedom and public health are incompatible; and so on across the spectrum of fears, denials, half-truths and indifference.

In this sense, epidemiology is as much an informational as an environmental science. In each case, the challenge is to understand and influence systems whose dynamics are best understood in terms of mutual dependency: as networks embodying certain propensities, feedback loops and tipping points. And while 'parasitical memes' may resemble microscopic organisms, the analogy with nature extends to other scales. Among other things, they resemble weeds, thriving opportunistically in environments too harsh for truth. Out there in the natural world, indifferent to the stories we tell, environmental degradation threatens us with a planet of weeds: choked lakes with few fish, oceans of algal blooms, forests of invasive pests. Online, wastelands inhospitable to complexity similarly expand: unquestioning monocultures of fear and anger, touched with righteous titillation.

In each case, exhortations towards personal responsibility, education and prevention only go so far. So long as the underlying vulnerabilities and imbalances remain unchanged, the crises they breed will also keep coming. Here, for example, is the author and technologist Jaron Lanier explaining how any information system designed to maximize engagement can end up snaring its users in a cognitive dead end as nothing more than the by-product of an optimization process.

> The algorithm is trying to capture the perfect parameters for manipulating a brain, while the brain, in order to seek out deeper meaning, is changing in response to the algorithm's experiments . . . As the algorithm tries to escape a rut, the human mind becomes stuck in one.[19]

It's dangerously easy, in the face of such accounts, to succumb to a pessimistic version of inevitablism: to flip from seeing human comprehension and intentions as supreme to dismissing them as illusory. The more important point, however, is that both understanding and agency remain eminently possible – but only if conceived as aspects of an ongoing, collective negotiation, obligated equally to address present knowledge and future uncertainties.

It's the story that makes the difference

In the larger scheme of things, both truth and human truth-seeking are by definition more resilient than denial can ever be. Ignoring reality may be a fine short-term strategy for demagogues or information constructs, but it's not a long-term recipe for survival – and it's this that is evolution's supreme diktat. A virus can kill you even if you refuse to believe it exists; the climate has no interest in narratives or factions.

What should we take from this? To echo the previous chapter's account of attentional ethics, any information environment aimed above all at emotional impact needs re-examination if we wish other strains of thought to remain vigorous. But this re-examination needs to be rooted in a rigorous account of *how* and *why* people can collectively arrive at better beliefs; and this means in turn that we cannot afford wishful thinking about rationality, persuasion or the force of facts. As Dennett notes:

> The notorious confirmation bias is our tendency to highlight the positive evidence for our current beliefs and theories while ignoring the negative evidence. This and other well-studied patterns of errors in human reasoning suggest that our skills were honed for taking sides, persuading others in debate, not necessarily getting things right.[20]

The question is not whether we can change – because, together, we are always in the process of becoming – but what the systems we are building enshrine and perpetuate. Are we playing a game in which truth and lies test their opposed strengths; or one in which winners get to make up the rules as they go along? Are we interested in putting our hopes and knowledge to a meaningful test; or would we rather fantasize life away rewriting history?

In each case, it's only through the principled embrace of doubt and uncertainty that we can prevent our longing for explanations – our ceaseless tidying of experience into human-scale stories – from tipping over into denial and delusion. And this embrace is in turn bound up with richer notions of narration: stories that can structure our hopes without trying to explain the world away; ways of resisting literal-mindedness that can equally harness our intellectual, imaginative and empathetic gifts.

In her 1986 essay 'The Carrier Bag Theory of Fiction', the novelist Ursula Le Guin draws a distinction between, on the one hand, accounts of prehistory that emphasize the spear as humanity's first defining tool and, on the other hand, accounts that emphasize the significance of carriers and containers: creations that support gathering, child-bearing, sharing and planning. It's a contrast that echoes the argument of this book's second chapter, as well as highlighting the moral force of the framings through which we attend to the world.

> It is a human thing to do to put something you want, because it's useful, edible, or beautiful, into a bag, or a basket, or a bit of rolled bark or leaf, or a net woven of your own hair, or what have you, and then take it home with you, home being another, larger kind of pouch or bag, a container for people, and then later on you take it out and eat it or share it or store it up . . . The trouble is, we've all let ourselves become part of the killer story, and so we may get finished along with it.[21]

It is, Le Guin argues, 'the story that makes the difference': the presumptions baked into our words and questions. Do we discuss technology in terms of competition, mastery and conquest; or in terms of culture, cooperation and common inheritance? Is our imagined future a battlefield, its territories warring and thickly walled; or is our planet a container whose contents we must jointly safeguard? There is no right answer, nor one inevitable fate waiting to be uncovered. Rather, the stories through which we frame our histories, opportunities and obligations are *themselves* among our most potent technologies. Hence this book: a container for words; a tale of imagined pasts and possible futures.

It's tempting, in the face of fake news, conspiracies and reality-denials, to double down on facts and evidence: to preach the gospel of rationality as our sole salvation. Certainly, civilization would benefit from an increase in reasoned self-examination. Yet when it comes to apprehending present and future challenges, the tools of myth, narrative, ambiguity and metaphor may be better able to bring us truths that *matter*: that help us grasp our interconnectedness via processes that are themselves endlessly open. As the psychologist Jonathan Haidt puts it in his 2012 book *The Righteous*

Mind, 'the human mind is a story processor, not a logic processor'[22]. Even the driest scientific paper is also a story about what counts, how it can be counted and why we should care; and assuming such things to be self-evident can be dismally self-defeating.

Ultimately, meaningful understanding and agency lie close to the heart of Virilio's paradox: in the iterative embrace of imperfectly understood consequences that are also always causes; in the collective business of attending to accidents, hidden histories and society's margins. Knowledge, if it is to be tempered by wisdom, demands newly ambitious forms of humility. It requires us to stop conflating technology with the tool-like extension of agency, and instead to conceive it as a collective, evolving site of values, identities and relationships. It asks us to confront the precipitous strangeness of existential risks, which can unmake millennia of progress in a moment. And it entails a rejection of the delusion that we can ever know in advance all that can or should be done: that comprehension is our entitlement, rather than a perpetual work-in-progress.

What do we need if we are to grow up as a species: to take possession of our potential while scrabbling for purchase on an exponential curve? The answer begins with an acknowledgement rather than a denial of technology's constant accidental revelations – and the embrace of narratives able to encompass these. As the philosopher Alasdair MacIntyre puts it in his 1981 book *After Virtue*, a line used as this book's second epigraph:

> I can only answer the question 'What am I to do?' if I can answer the prior question 'Of what story or stories do I find myself a part?'[23]

In the end, all of our hopes demand the same thing: that we find a way, together, to tell stories at once rich enough to reflect reality and resonant enough to command belief.

PART TWO

Destinations

CHAPTER 7

Values and assumptions:
The delusion of neutrality

In planetary terms, we stand at a threshold that's as straightforward to summarize as it is staggering to contemplate. We are entering an era in which we possess both the capacity to transform our world and the knowledge to do so deliberately, in the light of meaningful understanding. The aeons stretching ahead of us potentially outnumber all those we have left behind. Yet the window available for efficacious action is measured in decades rather than centuries: decades within which millennia of change will be set in motion.[1]

In effect, technological determinism is an overstatement of the self-evident truth that history, biology and physics constrain us. It confuses the fact that much is beyond our control with the conclusion that everything is: a defeatist misreading of history that can also become an all-too-convenient category error, implying every present act of profiteering or exploitation to be somehow preordained. Against this, the preceding chapters have suggested that an attentiveness to hidden histories and accidental revelations might keep us more honest, alongside the kind of incremental, iterative interventions fit for reality's complexities. We need richer stories through which to understand our possible futures – and the second half of this book aims to tell some of these.

To begin, this chapter addresses a claim closely connected to technological determinism: the belief that technologies are *neutral*, and that the question of what to do with them can thus be treated in isolation from questions like *how* and *why* they were made in the first place (not to mention *whether* they should exist at all). Once again, this line of argument is extremely convenient for those who would rather remain unaccountable. But it is also unfaithful to reality, for the human-made world is saturated with values and judgements. Indeed, its greatest challenges are fundamentally *ethical* in nature – and this in turn obliges us to address both the purposes threaded through a technologized age and what it might mean to alter them.

There are no neutral tools

To echo the American historian of technology Melvin Kranzberg, there is no such thing as a neutral tool.[2] Consider transport. To enter a vehicle is to transform your relationship with geography in particular ways. It is to take up a certain amount of space; to be able to move at a certain speed, with a certain carrying capacity; to be able to access certain areas; to take on particular power and maintenance requirements, served by particular infrastructure; to assume certain legal responsibilities. You can do some things in a vehicle, but not others. Using one grants options and freedoms that may be indispensable, or highly desirable; while lacking one in an environment expressly designed around their capabilities – to be unable to afford a car in Los Angeles, say – places you at the sharp end of a host of assumptions about freedom, space and society.

Here is a second, starker example. To pick up a weapon is to move through a world populated with potential targets. If I have a gun holstered on my belt, this changes me and my relationship with others in ways that can only be understood by analysing what the new entity 'me-and-my-gun' is capable of and disposed towards.

Similarly, to frame human struggles and nature in terms of firearm ownership is, in Ursula Le Guin's phrase, to give the 'killer story' pride of place when it comes to social and political systems: to suggest, for example, that the best protection against armed assault is to carry a gun yourself; or that owning a weapon is a natural and fundamental freedom.

In his 1992 essay 'Where Are the Missing Masses? The Sociology of a Few Mundane Artifacts', the philosopher Bruno Latour makes an intriguing point about these tendencies and assumptions.

> The distinctions between humans and nonhumans, embodied or disembodied skills, impersonation or 'machination', are less interesting than the complete chain along which competencies and actions are distributed.[3]

Although his prose can be oblique, Latour's image of a 'chain along which competencies and actions are distributed' is an elegant evocation of the technologized world's complexity: of a context within which both our *options* and our *intentions* are moulded by countless systems, tools and assumptions. We live amid a filigree of invisible chains. They constrain us, connect us, grant us agency. We can never be free of them. But we can notice, question and seek to alter the ways in which they are forged.

What follows from this? As slogans like 'guns don't kill people, people kill people'[4] suggest, the seductive notion that technology itself is neutral – that a tool is simply a tool, and all that matters is how it's used – is all too frequently evoked in order to evade discussion of the assumptions and possibilities it embodies, not to mention the value-laden systems of regulation, power and profit surrounding it.

If technologies themselves are neutral, the people who make and maintain them have no responsibility towards the people who use them (and upon whom they're used) beyond ensuring certain standards of quality and functionality. If the most one can say about a town in which everyone walks around holding an assault rifle is that it's up to them to use their rifle responsibly, the question of what it means to live in a community where lethal force is a constantly visible prospect makes no sense. All that can be expressed is a hope that people use their military-grade weapons 'well' – whatever that might mean in the context of an artefact designed expressly to kill in combat.

To talk about the possibilities, values and preferences instantiated in technologies is to talk about what are sometimes called their *affordances*: a term coined by the psychologist James J. Gibson in 1966[5] and most influentially articulated in his 1979 book *The Ecological Approach to Visual Perception*, where he defined it in terms of the complementary relationship between an animal and the environmental 'offerings' it is able to use to its advantage.

> The affordances of the environment are what it offers the animal, what it provides or furnishes, either for good or ill. The verb to afford is found in the dictionary, the noun affordance is not. I have made it up. I mean by it something that refers to both the environment and the animal in a way that no existing term does. It implies the complementarity of the animal and the environment.[6]

Humans, Gibson emphasizes, are much like other animals when it comes to their environment's most fundamental affordances: shelter and sustenance drawn from the oceans, air and soil; the cycles of water, nutrition and energy that sustain and replenish them. We are prodigiously talented at spotting, exploiting and creating affordances: indeed, one way of telling technology's story is via its transformation of our surroundings into hyper-providers of comfort, security and agency. Yet none of this means that we occupy a different world to every other creature. As Gibson notes:

> This is not a new environment – an artificial environment distinct from the natural environment – but the same old environment modified by man. It is a mistake to separate the natural from the artificial as if there were two environments; artifacts have to be manufactured from natural substances. It is also a mistake to separate the cultural environment from the natural environment, as if there were a world of mental products distinct from physical products. There is only one world, however diverse, and all animals live in it, although we humans have altered it to suit ourselves.[7]

Gibson's argument is significant both because of its insistence that 'there is only one world' and because of its implication that we can only fully assess any technology's offerings by looking at both what it *gives* and what it *takes*: what it facilitates and what it effaces. To borrow a phrase from the author Kevin Kelly, it is our duty to ask 'what technology wants' – and to do so in the knowledge that the answer may entail assumptions, preferences and inclinations at odds with our own. In what sense can a technology be said to 'want' anything? Here's Kelly writing in 2010 about what he means by the phrase.

> [Technology] wants what we design it to want and what we try to direct it to do. But in addition to those drives, [it] has its own wants. It wants to sort itself out, to self-assemble into hierarchical levels . . . [it] wants what every living system wants: to perpetuate itself, to keep itself going. And as it grows, those inherent wants are gaining in complexity and force. I know this claim sounds strange. It seems to anthropomorphise stuff that is clearly not human. How can a toaster want? . . . But 'want' is not just for

humans . . . mechanical wants are not carefully considered deliberations but rather tendencies. Leanings. Urges. Trajectories.[8]

Much as they elude complete control and comprehension, technological systems generate their own forms of momentum; and these embody powerful forms of 'wanting' in much the same sense as water can 'want' to pour through the cracks in a damaged cup, or a marble can 'want' to roll down the grooves in a children's toy.[9] To return to this chapter's opening example, a car wants me to move at speed along tarmac roads in a private metal box. Or, more precisely, the car-and-I want this when we are acting in concert. This is a central aspect of its offering. And if I and a sufficient number of others accept it – or have little choice but to accept it – our towns and cities will start to be planned around such affordances. Gradually, they will come to afford greater convenience and control to those who navigate them within cars and less to those who do not, unless other priorities and technologies are borne in mind. An implicit negotiation plays out, in other words, every time a design decision is made or a technology adopted. And this applies *especially* in situations where most people are unaware it's happening.

Consider some of the simplest technologies in my own life: those surrounding me as I type these words. The chair in my office wants me to sit. Different kinds of chair want me to sit in different ways, but all of them offer an elevated form of comfort compared to sitting on the ground. The fact that for much of the day I sit in a chair reading, writing and interacting with a computer has considerable implications for my health and mental state. Not all of these are positive; but I'm immensely grateful not to be undertaking back-breaking labour in a field. As Douglas Adams noted, chairs were once uncommon and desks still more so. Today, I spend much of my time happily fixated on a screen's simulations and summonings of people and places. And this is all very well, so long as I retain

VALUES AND ASSUMPTIONS

some awareness of (and control over) the interplay between what is offered and what may be being withheld.[10]

All of this begs a twinned set of questions when it comes to any and every aspect of the human-made world. What do I want; and what do the systems I'm enmeshed in want from me in turn? When it comes to a technology like email, I'm conscious that my inbox would like me to spend every single minute of every single day emptying it, all the while filling up the inboxes of everybody else I know. For many people, an email inbox is in effect a to-do list written for them by other people – a vital aspect of work and life that can also become an unending and resented source of labour. Why? Because sending emails is instantaneous and costs nothing but time; because it serves as proof that you're working and attending to certain tasks; because it's useful; and because many people have access to their inboxes every waking moment of every single day.

Perhaps most importantly, an email 'wants' you to send another email rather than to pick up the phone or write a letter, just as an always-on mobile device 'wants' you to attend to it rather than to your environment or the people beside you. As we saw in the previous chapter, your time and attention are ceaselessly being shaped by the artefacts surrounding you – and this means that *you* are too. Many of technology's wants are unintended: accidental revelations, in Virilio's phrase, born from a combination of coincidence and momentum. But that doesn't make them any less forceful, or the prize of a meaningful negotiation with them any less precious.

When it comes to technologies like social media, moreover, both a system's and its designers' desires may be more insistent – and married to outcomes that its users are, at best, ambivalent towards. Even conceiving of someone's relationship with a technology as that of a 'user' is charged with particular assumptions and implications. Consider what it means, by contrast, for a technology to be designed and debated with the needs of a 'citizen', a 'creator' or a 'collaborator' in mind. And this is before you get to areas such as surveillance, social engineering and restrictions of liberty, where technology can embed coercion into the very fabric of an environment: where the affordances of the human-made world may entail not so much a negotiation as an attempt to eliminate the very possibility of dissent. This is the point at which determinism comes closest to being true – at least until we discern, behind each precision-engineered surface, the presence of human desires and decisions. The *delusion of neutrality* itself is, in other words, far from neutral in ethical terms.

Where determinism meets neutrality

What are the most significant ways in which present technologies differ from those that came before? In her 2016 book *Technology and the Virtues*, the philosopher Shannon Vallor makes the case that their power and sophistication have brought into being a global, networked moral landscape.

> The invention of the bow and arrow afforded us the possibility of killing an animal from a safe distance – or doing the same for a human rival, a new affordance that changed the social and moral landscape. Today's technologies open their own new social and moral possibilities for action. Indeed, human technological activity has now begun to reshape the very planetary conditions that make life possible . . . our aggregated moral choices in technological contexts routinely impact the well-being of people on the other side of the planet, a staggering number of other species, and whole generations not yet born. Meanwhile, it is increasingly less clear how much of the future moral labour of our species will be performed by human individuals.[11]

VALUES AND ASSUMPTIONS

As Vallor's account makes clear, an almost unthinkably consequential degree of complexity is inherent to the twenty-first century's interconnected technologies. And this makes exploring the gestalt nature of modernity's 'moral labour' – its diffusion of responsibility between those designing, making, regulating, using and profiting from different technologies – a vital corrective both to the delusion of technological neutrality and its entanglement with deterministic accounts of innovation.

Let us return, for a moment, to the fundamentals of technological determinism. Many different versions of its claims exist, but underlying them all is the belief that new technologies bring with them behaviours and outcomes that, over time, define the course of history.[12] Thus, to borrow a phrase from another philosopher of technology, L. M. Sacasas, 'resistance is futile' when it comes to challenging these outcomes.

Sacasas himself borrows the phrase from no less an authority than *Star Trek: The Next Generation*, where it's the battle cry of the Borg Collective, a cyborg civilization whose mission is to assimilate all other life-forms into their hive mind. 'Resistance is futile!' its drones repeat as they try to extinguish every form of freedom alien to their own.[13] They're wrong, of course: the Star Trek universe wouldn't be much fun if resistance was indeed futile. But their sinister hubris is a handy – and gloriously heavy-handed – metaphor for all those mindsets that insist upon technology as a form of destiny. As Sacasas notes, to identify and oppose what he calls the 'Borg Complex' mode of tech analysis is to assert the ethical significance of taking responsibility for our creations.

> Marshall McLuhan once said, 'There is absolutely no inevitability as long as there is a willingness to contemplate what is happening'. The handwaving rhetoric that I've called a Borg Complex is resolutely opposed to just such contemplation when it comes to

technology and its consequences. We need more thinking, not less, and Borg Complex rhetoric is typically deployed to stop rather than advance discussion. What's more, Borg Complex rhetoric also amounts to a refusal of responsibility. We cannot, after all, be held responsible for what is inevitable.[14]

At this point, it's worth emphasizing one of the strangest features of claims of technological neutrality and inevitability: that, even though they directly contradict one another, they're often articulated together.

To say that a tool is neutral is to say that its users bear responsibility for what's done with it, presumably on the basis that this is their free choice. By contrast, to say that technology has an internal logic dictating certain outcomes is to say that people cannot ultimately choose whether or how to use it: dissent is the province of Luddite fools. Yet this deterministic rhetoric often dovetails with rhapsodies upon user empowerment. As the CEO of Evernote, Phil Libin, put it back in 2012 when discussing the then-new (and subsequently discontinued) augmented reality Google Glass:

> I'm actually very optimistic about the Google Glasses – and those by other companies who will make them . . . I've used it a little bit myself and – I'm making a firm prediction – in as little as three years from now I am not going to be looking out at the world with glasses that don't have augmented information on them. It's going to seem barbaric to not have that stuff. That's going to be the universal use case. It's going to be mainstream. People think it looks kind of dorky right now but the experience is so powerful that you feel stupid as soon as you take the glasses off . . . [15]

It's all too easy to play the game of digging up predictions that didn't come true (more recently, the Metaverse spent eighteen months being The Future before it was quietly dumped).[16] But what's telling about Libin's line of argument is its treatment of human desire and technological possibility as two sides of the same coin. Google Glass offers such a great experience that anyone who uses it will, seemingly inevitably, want to keep on using it. To do otherwise will soon become 'barbaric': it will mean existing outside the grand progress of technological civilization.

In the best pseudo-Darwinian style, this framing suggests that technology's powers sooner or later make its offerings synonymous with the outcome of a free choice; and that such choice is thus an illusion when it comes to aggregated human behaviours over time. Apparently, people are being gifted more opportunities than ever before by products and platforms whose dominance is preordained: a reading of history that's only plausible if you ignore the chaotically branching possibilities, debates, rethinks and repercussions surrounding every innovation.

This brings us to the heart of an ethical concern that recurs throughout the second half of this book. Delusions like neutrality and inevitability matter not only because they deny the existence of any choices more fundamental than 'which app shall I install next?' but also because, by doing so, they negate any basis for a discussion of what we *ought* to aspire towards that isn't based upon either expert condescension (*please invent the great innovation that will inexorably save us!*) or the decontextualized praise of individual responsibility (*please use your assault rifle responsibly!*).

In each case, what purports to be ethical engagement is little more than wishful buck-passing: the pretence that we live in a world where the complexities of our 'aggregated moral choices in technological contexts' can be palmed off as non-issues or personal preferences. What's the alternative? As ever, it begins with paying close attention to what's *actually* going on.

The assumptions embedded in technologies

Near the start of their 2018 book *Re-Engineering Humanity*, law professor Brett Frischmann and philosopher Evan Selinger explore an example of what they term 'techno-social engineering' at Oral Roberts University in Tulsa, Oklahoma. In 2016, the university introduced a requirement for students to purchase and wear Fitbit tracking devices for a physical education class. Previously, students had self-recorded their daily activities in a journal. Now, these activities would automatically be recorded by their devices.

A minor controversy ensued concerning how far students had given informed consent to this tracking, how data would be stored, and so on. But this controversy faded once it became clear that the university had provided adequate safeguards. One kind of monitoring had simply been replaced by another: the technology of pens and paper by automated tracking. Who, in this day and age, would seriously suggest things should be different? Indeed, who would deny that Fitbits provide more detailed and more reliable data than journals, and do so more conveniently?

Frischmann and Selinger aren't in the business of mourning pens and paper. But, by digging into the different affordances of old and new approaches, they unearth some significant complexities. For a start, they suggest, there are profound psychological differences between actively recording observations and passively being monitored.

> Students who record their daily physical activities in a journal find the analog medium affords several steps that require time and effort, planning and thinking. It can orientate students to record fitness data in ways that automated and unreflective inscription machines could never do. The medium directs student attention

inwardly and outwardly and the recorded data can reveal more than meets the eye.[17]

For Frischmann and Selinger, it's this active/passive distinction, not the presence or absence of any particular technology, that matters. What's at stake is a certain ethic or set of values.

> Think-and-record activities inspire self-reflection, interpersonal awareness, and judgement. These activities are valuable because they're linked to the exercise of free will and autonomy . . . The key to techno-social engineering better humans just might lie in taking these slower tools more seriously.[18]

Within the space of two paragraphs, we have moved from a description of students scribbling in journals to a discussion of values associated with being a 'better' human being. Is this move justified? The answer is an emphatic *yes* – and one that's all the more important for the starkness of placing such an ethically charged claim alongside what might more often be treated as a minor example of tech-enabled efficiency.

To see why, we need to consider not only students' actions and options, but also the obligations and expectations accompanying them. To ask someone to use a wearable device is to ask them to consent to a process of observation that will automatically generate exhaustive data about their daily activities. By contrast, asking them to record their own actions means asking them to embark upon a process of self-observation – and trusting them to do so diligently. This second scenario requires not only practical effort but also the kind of moral labour highlighted by Vallor: undertaking to perform a task accurately and honestly while resisting the temptation to distort or fabricate its results.

Especially in the context of education, it's reasonable to ask what kind of a student each of these approaches *wants* someone to be – and what standards it suggests they'll be assessed by. Is a good student someone who can be trusted to take responsibility for a sustained self-assessment; or is it someone whose comfort and convenience are best served by unobtrusive automatic monitoring? You might reply that the most realistic answer is 'a bit of both' – but it's not obvious that both options are on offer.

The implications of choices like this extend well beyond their immediate context. What kind of a person are students being encouraged to grow into by an education system that assumes constant, automated monitoring is a necessary feature of the world? What might it mean for a society to integrate such surveillance into the fabric of education; for students to perform all their schoolwork on devices that automatically report on their actions or inactions; for facial recognition systems to track attentiveness in classrooms in real time?

None of these scenarios are hypothetical. Here's how Todd Feathers and Janus Rose reported for *Vice* magazine's Motherboard website in September 2020 on the growing use of 'digital proctoring' software to monitor students in some US colleges.

> The software turns students' computers into powerful invigilators – webcams monitor eye and head movements, microphones record noise in the room, and algorithms log how often a test taker moves their mouse, scrolls up and down on a page, and pushes keys. The software flags any behaviour its algorithm deems suspicious for later viewing by the class instructor.[19]

Dystopian though it may sound, there are clear reasons for the widespread adoption of such tools. The Covid-19 pandemic and its aftermath have led to rapid increases in remote learning and assessment. This has in turn left colleges struggling with what it means to monitor students working from home, as well as preventing plagiarism and cheating on a mass scale while coming up with measurable proxies for attendance and participation.

So long as software is deployed responsibly, you might say, this is all very well: surely the diligent and the innocent have nothing to fear? As Motherboard's account suggests, however, this defence starts to founder once the affordances of remote technologies are more closely scrutinized. In the case of proctoring software designed to monitor online exam-taking, for example, a factor that should be entirely irrelevant to any assessment – the colour of someone's skin – can become a major obstacle thanks to the fact that some facial recognition systems repeatedly classify those with darker skin as being too poorly lit to recognize. Similarly, students with unreliable internet connections, disabilities, anxiety, ADHD, or who live in close quarters with dependants, are more likely to be flagged up as 'suspicious' thanks to the patterns of their gaze, their keyboard and mouse use, their physical environment, their logon timings, and so on.

In such cases, the assumptions imparted to automated systems about what is desirable and 'normal' can't be separated from larger questions about the nature of twenty-first-century education, or indeed about membership in a twenty-first-century society. As its vendors have pointed out, colleges are under no obligation to use software in any particular way, or indeed at all. But its very existence embodies a powerful set of incentives and assumptions around trust, privacy and success as a student in the twenty-first century. And – crucially – it's not the only model out there, either for education or technology. Alternative approaches and attitudes exist; and many students and educators have expended a great deal of effort asserting their ethical and practical superiority.[20] As the educationalist Sioux McKenna noted in August 2022, the very decision to treat students as would-be cheats who need to be minutely scrutinized while regurgitating chunks of information is a diminishing one.

> When universities become businesses selling qualifications, it narrows their potential to be places where students enjoy transformative relationships with knowledge, and where knowledge is created to serve people and the planet . . . What most students need is to understand how knowledge is made in their field of study, what contributions that field makes to society, and how they can source and evaluate information to answer questions and resolve problems. They need to learn how to be ethical, critical citizens.[21]

These issues are all the more important in the context of AIs able endlessly to produce essays and answers, not to mention the crises such systems have provoked around assessment and originality. One response is, in effect, to let the 'killer' story run rampant: to reimagine education as an arms race between warring sides. More compelling and sustainable, however, are strategies that ask what it means to use novel technologies *well* – and for students to acquire meaningful knowledge and understanding in an era of ubiquitous algorithmic mediation.

The disruption of existing models and approaches, in other words, only makes an interrogation of the values embedded within emerging systems *more* urgent. Even if surveillance can be made to work effectively and impartially, what does it mean for a society to make submission to such monitoring a model for education, employment and civic life? As Evan Selinger and the philosopher Evan Greer put it in a February 2020 article, warning against the move to deploy facial recognition technologies on university campuses:

> Given the many ways [on-campus facial recognition] technology can be used and the ease of adding its functions to existing

cameras, any deployment will normalize the practice of handing our sensitive biometric information over to private institutions just to get an education . . . Indeed, the mere prospect of widespread facial surveillance will have a chilling effect on campus expression. Students who are afraid to be themselves and express themselves will pull back from crucial opportunities to experience intellectual growth and self-development – and students from marginalized communities will be the most affected.[22]

Societally, such software is of a piece with moves under way everywhere from business and leisure to governance and administration: towards the normalization of surveillance and algorithmic data-processing in the name of security and convenience; towards offers of efficiency and simplicity behind which under-examined prejudices or explicitly exploitative motives may lurk; and towards a fundamental asymmetry between what users understand versus what others understand about them. Indeed, the prospect of entire nations introducing regimes of total technocratic surveillance is today not so much speculative fiction as increasingly well-documented reality.

Frischmann and Selinger touch on all of these concerns in *Re-Engineering Humanity*. Yet they don't end their opening chapter with a jeremiad. Instead, having analysed the affordances of old and new approaches in the case of Oral Roberts University, they suggest some modest positive steps that might be taken based on such an analysis:

> The university could combine the fitness tracking tools. It could require students to use a fitness tracking device that collects data, while also expecting them to write reports about the collected data in a journal. This two-step process would be more comprehensive and accurate than journaling alone. It also gives students an

opportunity to reflect on their performance and freedom to define how and what to communicate to their instructors and peers. As a deeper exercise, students might be asked to reflect on the data, what it says and doesn't say about them.[23]

Once the right questions have been asked, a negotiation can in principle take place between different systems and approaches, animated by a clear discussion of what human ends the result should be directed towards – and what might need to be mitigated along the way. The right questions can only begin to be asked, however, if technology's affordances are borne in mind and the delusion of its neutrality dispensed with. What's required is an explicitly ethical understanding of the assumptions embodied in a technology's design and deployment – and the permission and the will to turn this understanding into action.

CHAPTER 8

Myths and wish-fulfilment:
The delusion of magical thinking

In December 2003, the author Joan Didion was having dinner with her husband of forty years, the author John Gregory Dunne, when he collapsed and died from a massive heart attack. It happened halfway through a conversation, in their New York apartment, while he was seated at the table and she was mixing a salad. He slid to the floor, an ambulance came, but nothing could be done.

Two years later, in her memoir *The Year of Magical Thinking*, Didion addressed the grief and trauma of her loss – and the strangeness of the 'magical' frame of mind that defined it. Perhaps because of the utter ordinariness of the events preceding Dunne's death, she struggled to make herself believe that it really had happened. This is what her book's title refers to: the irrational yet irresistible hope that a broken reality might be mended by sufficient belief. 'I was', Didion writes:

> thinking as small children think, as if my thoughts or wishes had the power to reverse the narrative, change the outcome.[1]

Throughout the book, Didion reveals two versions of herself. There is the author, an unflinchingly clear-eyed observer of her own and others' lives; and there is the grief-struck widow, haunted and bewildered, struggling to grasp the altered facts of her existence. She finds that she cannot throw away Dunne's shoes or clothes because, a part of her whispers, he's going to need them when he comes back. This cannot be true, and she knows it. Nevertheless, she feels that it could *become* true, if only the magic of belief were sufficiently mobilized. Her grief is not just a matter of suffering and mourning. It is also an unanticipated departure from reality's old coherence into a new, strange story.

> Grief turns out to be a place none of us know until we reach it . . . We might expect that we will be prostrate, inconsolable, crazy with loss. We do not expect to be literally crazy, cool customers who believe that their husband is about to return and need his shoes.[2]

Surely, Didion thinks, no modern and reasonable person should expect a grief like this. Yet it is hers. For eleven months, she lives in a place of rituals, superstitions, denials and fantasies; of flights of delusion in the face of unbearable absence. Finally, these start to recede as she undertakes an arduous reconstruction of the events of that evening – and tries to accept that they were not her fault. There is, in a sense, nothing personal about this most personal of losses. There is just the universal fact of human fragility, against which magical thinking offers a retreat into stories of remedy, transformation, hope and blame:

> I realize how open we are to the persistent message that we can avert death. And to its punitive correlative, the message that if death catches us we have only ourselves to blame.[3]

We all think wishfully at times. All of our thoughts are touched by magic, in that we long for feelings and fantasies to become facts about the world. But this is tempered by the insistent unfolding of events – and by our capacity, collectively, to confront painful truths and update our models of reality. In this sense, magical thinking is not so much a discrete mental state as an extreme indulgence of the inner at the expense of the outer: a literal reading of things that cannot literally be true.

How does all this relate to technology? The previous chapter explored how preferences and potentials are embedded in all technologies: how they *afford* certain forms of thought and action while effacing others. What this can only hint at, however, is the ways in which the stories we weave around technology can take on a life of their own; and how wishful thinking can recast our creations as literal embodiments of purpose, destiny and even salvation.

Faith in technology as a new religion

In her 1966 novel *Les Belles Images*, the philosopher Simone de Beauvoir imagines the life of a successful but unfulfilled woman working in Paris's advertising industry. Here is how her protagonist, Laurence, thinks about the stories propagated by advertisers through the parallel realms of production and consumption:

> Smoothness, brilliance, shine; the dream of gliding, of icy perfection . . . To astonish and at the same time to reassure; behold the magic product that will completely change our lives without putting us out in the very least.[4]

It is the yearning for this 'magic product' that, de Beauvoir suggests, we must excise if we are to escape the pursuit of mass-manufactured, inhuman perfection – not to mention the covert narratives of insecurity and inadequacy that fuel it.

To remain in the Francophone world for a moment, consider another prophetic mid-twentieth-century work: the French philosopher and theologian Jacques Ellul's 1954 book *The Technological Society*, parts of which anticipate Kevin Kelly's account of technology's 'wants'. In *The Technological Society*, Ellul explores a version of magical thinking characterized by the ways in which technical achievements can become objects of reverence based on nothing more than ingenuity and novelty. Among other examples, he considers the hydroelectric dams built and managed by the Tennessee Valley Authority (TVA) after it was set up by Congress in 1933 in order to generate electricity and prevent flooding. The initial engineering phases of the project were successfully and profitably undertaken: power distributed locally, the landscape altered, flood risks reduced. From the technical perspective, certain material benefits were delivered via a certain amount of ingenuity and effort. But, Ellul notes, such a description barely scratches the surface of what the TVA came to mean:

> even before the program yielded concrete results, the myth began to develop, and today the TVA has become a symbol of regionalism in the United States. To it is ascribed the function of co-ordinating and integrating diverse activities; a role in the methodical development of natural resources; a task of decentralization affecting public and private federal and local institutions; and even a mission of education . . . It is literally impossible for the public to believe that so much effort and intelligence, so many dazzling results, produce only material effects. People simply cannot admit that a great dam produces nothing but electricity . . . In short, man creates for himself a new religion of a rational and technical order to justify his work and to be justified in it.[5]

This 'new religion of a rational and technical order' embodies a fundamentally circular form of reasoning. Once enough effort is expended, any sufficiently dazzling technical achievement can become its own justification. As Ellul also noted:

> the feeling of the sacred is expressed in this marvellous instrument of the power instinct which is always joined to mystery and magic. The worker brags about his job because it offers him joyous confirmation of his superiority. The young snob speeds along at 100 m.p.h. in his Porsche. The technician contemplates with satisfaction the gradients of his charts, no matter what their reference is. For these men, technique is in every way sacred; it is the common expression of human power without which they would find themselves poor, alone, naked, and stripped of all pretentions. They would no longer be the heroes, geniuses, or archangels which a motor permits them to be at little expense.[6]

While it's possible to take issue with some of Ellul's analyses (he underplays, for example, the explicitly ideological origins and framing of the 1933 Tennessee Valley Authority Act, and how the TVA was as much an expression as a cause of certain beliefs),[7] he is brilliantly sensitive to the unearned faith our creations can inspire through their promise of purchased superiority: through the ceaselessly marketed pursuit of powers which can be ours at little (or, better yet, considerable) expense.

The problem isn't that all technical achievements are to be deplored. Rather, it's that they must be judged by criteria beyond speed, power and scope. As the author Evgeny Morozov similarly argues in his 2013 book *To Save Everything, Click Here*, it's tempting to treat the world as a series of problems awaiting technological solutions, especially if you're in the business of selling such solutions. But this approach shirks the kind of careful, attentive investigations demanded by any complex social challenge.

> Recasting all complex social situations either as neatly defined problems with definite, computable solutions or as transparent and self-evident processes that can be easily optimized . . . presumes rather than investigates the problems that it is trying to solve, reaching [in the words of design theorist Michael Dobbins] 'for the answer before the questions have been fully asked.'[8]

For de Beauvoir, Ellul and Morozov alike, thinking magically about technology is toxic because it encourages us to worship innovation and consumption as ends in themselves – while submitting to the story that some 'magic product' may solve all our problems 'without putting us out in the very least.' It's a belief that lurks beneath the surface of countless publicity campaigns and perkily soundtracked adverts, often framed in terms of an autonomous, enlightened *you* whose life will inexorably be enhanced by mechanical marvels. *You* sit down and switch your laptop on; *you* slip into your oh-so-smart car; *you* reach for your phone. 'What do *you* want to do today?'[9] asks the waiting software. 'What do *you* want to know, or do, or buy, or consume?' The second person singular is everywhere. *You* are empowered, *you* are enhanced, *your* mind and body extended in scope and power. All technology can be judged by how fast it allows *you* to dash in pursuit of desire. Like Icarus unencumbered by frailty, you'll ascend forever into cloudless skies.

This is mystery and magic reborn as marketing: the claim that, at least for those with adequate funds, technological progress offers an infinitely accommodating form of empowerment. As another twentieth-century philosopher and theologian, Ivan Illich, put it in his 1978 book *The Right to Useful Unemployment*, the right to consume innovation's fruits is in this world-view a panacea – a birthright from which all good things flow:

> An ever-growing part of our major institutions' functions is the cultivation and maintenance of three sets of illusions which turn the citizen into a client to be saved by experts . . . The first enslaving illusion is the idea that people are born to be consumers and that they can attain any of their goals by purchasing goods and services.[10]

The second 'enslaving illusion', Illich goes on to suggest, is that technological progress demands ever more expert control at the expense of wider participation; while the third is that 'good things will forever be replaced by better things'. In each case, what's 'enslaving' about these attitudes is their transfer of power and responsibility away from individuals towards a technocratic elite entrusted with 'solving' the world on everyone else's behalf. This is the *delusion of magical thinking* writ large: a wishful abnegation of responsibility in favour of technologized salvation. And it's also an approach that is unable to deliver lasting solutions even on its own terms. As Morozov puts it at the end of *To Save Everything, Click Here*:

> it would be ironic if humanity were to die in the crossfire as its problem solvers attempted to transport that very humanity to a trouble-free world.[11]

Trouble-free worlds do not exist except in fairy tales. And the best antidote to them isn't yet more wishful thinking. It's the telling of wiser tales: ones that can help us *stop* pretending the world is a collection of problems awaiting technocratic solutions.

Moving beyond pretence

How can we tell stories about technology that don't conflate archetypes, propaganda and ad campaigns with real life? It begins with a retreat from wish-fulfilment in all its forms – and with the rejection of notionally autonomous, individual consumers as the sole protagonists of life's story.

The modern world is a network of networks. It girdles the human, beneath and beyond our attention, connecting not only person to person but also object to object through petabytes of shifting data. It bristles with opportunities for the enhancement of human freedom and agency. But the kind of cost-free individualism depicted in marketing and media is invariably a consumerist category error – while the 'you' that matters most is, like its machine extensions, strictly plural. As the author and political scientist Thomas Rid puts it in his 2016 book *Rise of the Machines*:

> Gaining control through machines means also delegating it to machines. Using the tool means trusting the tool . . . The connective tissue of entire communities has become mechanized. Apparatuses aren't simply extensions of our muscles and brains; they are extensions of our relationships to others – family, friends, colleagues, and compatriots.[12]

For Rid, the challenge is not only to address the extension of human minds and capabilities through technology, but also the extension and interlocking of *many* minds: of selves interconnected to an unprecedented degree, both through and with machines. We aren't in competition with our creations. Rather, they are the stuff we are made of: a language, a culture, a looping feedback between all that humanity has made but did not choose.

Facing up to this means both finding new ways of talking about technology's promises and rediscovering old ones. Unlike the wishful literalism of magical thinking, the best and most powerful stories take on new forms in the light of new knowledge. They fortify us precisely because they *refuse* the pretence that life can be entirely understood or anticipated: that dogmatism or denial are ever enough. What we need is to replace the false comforts of magical thinking with stories that are alert to the *mythic* in a very different sense: one that can give structure to our hopes, fears and longings without offering false consolation.

In her 1976 essay 'Myth and Archetype in Science Fiction', written a decade before 'The Carrier Bag Theory of Fiction', Ursula Le Guin makes an impassioned case for the significance of this kind of story. To dismiss myths as merely regressive fantasies is misguided, Le Guin suggests, because this misses much of what matters about human apprehensions and experiences.

> Myth is an expression of one of the several ways the human being, body/psyche, perceives, understands, and relates to the world. Like science, it is a product of a basic human mode of apprehension . . . We are rational beings, but we are also sensual, emotional, appetitive, ethical beings, driven by needs and reaching out for satisfactions which the intellect alone cannot provide. Where these other modes of being and doing are inadequate, the intellect should prevail. Where the intellect fails, and must always fail, unless we become disembodied bubbles, then one of the other modes must take over. The myth, mythological insight, is one of these. Supremely effective in its area of function, it needs no replacement.[13]

Ultimately, Le Guin suggests, mythological insight's 'area of function' is the question of *who* and *what* we believe ourselves to be: individually, collectively, as part of life's unfolding pattern. As she put it in a 2005 talk at the Conference on Literature and Ecology in Eugene, Oregon:

> The general purpose of a myth is to tell us who we are – who we are as a people. Mythic narrative affirms our community and our responsibilities . . . Fearful and suspicious as it is, yet the human mind yearns towards a greater belonging, a vaster identification.[14]

We may not always notice that such narratives run beneath the surface of our lives. But they are there whenever we ask what binds and divides us; what we owe to one another; what it means to invoke nations, histories, races, cultures, faiths. How can we hope to live and to die well; to map the past and plan for the future? Which injustices demand redress, which triumphs celebration? These are all questions that can – and should – be elucidated by reasoned investigation. But they cannot be exhausted by it, for there *is* no final resolution to be had: no last word to be spoken, no technical achievement poised to dissolve them. In the end, it is how we inhabit and interrogate these mysteries that matters: how far the stories through which we understand ourselves illuminate their threats, promises and contradictions. As Le Guin puts it: 'The real mystery is not destroyed by reason. The fake one is.'[15]

Harnessing myth and mystery

Below are two of my favourite ancient myths, followed by one of their most iconic modern incarnations. Among other things, they're about hubris, usurpation and seduction – and how all three continue to echo beneath our best achievements.

In Greek mythology, the sculptor Pygmalion falls in love with his supremely beautiful creation, Galatea. In the Roman poet Ovid's telling, there's a happy ending. Venus, the goddess of love and counterpart to ancient Greece's Aphrodite, takes pity on Pygmalion and breathes life into the marble. The statue's lips grow warm under his kiss. It's a moment of sensual revelation that is at once glorious and disconcertingly one-sided: the sculptor bestowing love upon his own fantasy made flesh. Here is the moment Galatea wakes to her maker's unrequested kiss, in Henry Riley's classic translation:

> She seems to grow warm. Again he applies his mouth; with his hands, too, he feels her breast. The pressed ivory becomes soft, and losing its hardness, yields to the fingers, and gives way ... at length [he] presses lips, no longer fictitious, with his own lips. The maiden, too, feels the kisses given her, and blushes; and raising her timorous eyes towards the light of day, she sees at once her lover and the heavens.[16]

Pygmalion and Galatea, in most versions of the story, live happily ever after. They fall in love, marry, have a child. But their tale has an unhappier classical cousin: that of Talos, the artificial man. Created by the divine smith, Hephaestus, Talos is often depicted as a bronze giant striding through the seas. Immensely strong, almost invulnerable, Talos renders all human might redundant. One of his most disturbing appearances is in the Elizabethan poet Edmund Spenser's 1596 epic *The Faerie Queene*, where 'Talus' is an iron squire committed to defending Justice alongside the good knight Sir Artegal. Here is how Talus deals with one of the obstacles they encounter, the defenceless daughter of a lord who has unjustly been filling his treasury with spoils from vanquished knights:

> Yet for no pity would he change the course
> Of Justice, which in Talus' hand did lie;
> Who rudely hauled her forth without remorse,
> Still holding up her suppliant hands on high,
> And kneeling at his feet submissively.
> But he her suppliant hands, those hands of gold,
> And each her feet, those feet of silver trye,

> Which sought unrighteousness, and justice sold,
> Chopped off, and nailed on high, that all might them behold.[17]

And this is just for starters. Having chopped off the lady's hands and feet and nailed them 'on high', Talus picks her up and throws her over the battlements of the castle before razing it to the ground. Even Artegal is disconcerted; but the iron man's enforcement of Justice, once set in motion, cannot be stayed. Mercy is not in the machine's repertoire.

Skip forward half a millennium and we find Galatea and Talos dovetailing into one of the 1990s' most iconic science fiction films, *Terminator 2: Judgment Day*, James Cameron's masterpiece of action and exquisitely honed musculature. *Terminator 2* has become a dystopian point of reference for debating Artificial Intelligence, in the process inspiring countless dubious front pages, headlines and begged questions. More intriguing and prophetic than its pyrotechnics, however, is a quiet moment during the second half of the film when Arnold Schwarzenegger's titular Terminator – an artificial killer reprogrammed to act as the perfect protector – is hanging out with his young protectee, John Connor.

The scene offers a momentary respite from pursuit. John's mother, Sarah, watches from a distance as the cyborg plays with the ten-year-old. Arnie has flipped from one polarity to the other; from perfect assassin to perfect playmate. And this prompts a revelation. 'It was suddenly so clear,' Sarah says in voiceover:

> The Terminator would never stop. It would never leave him, and it would never hurt him, never shout at him, or get drunk and hit him, or say it was too busy to spend time with him. It would always be there. Of all the would-be fathers who came and went over the years, this thing, this machine was the only one that measured up.[18]

Tireless, infinitely patient, endlessly consistent, our creations 'measure up' in ways we can only dream of. Who wouldn't want an immaculate machine companion, employee, parent, lover? Then again, what does it mean to want such things: to be willingly seduced by artifice?

The embrace of manufactured perfection in place of fleshy vulnerability is a theme older than Ovid. As elements of it start to cross over into reality, however, there's a fresh urgency to its tensions – and this is exactly as it should be. Myths are there, among other things, to warn us about the weakness of human desire and judgement. To become entirely human, as in Pygmalion's happy ending, is one thing. But to supplant the human is quite another.

I love *Terminator 2* both because it's overtly a fiction – a story that demands we *don't* take it literally – and because it so deftly handles its ambiguities. Arnie is there to help humans do human things: save the world, blow stuff up, chase around in trucks and motorbikes. But it's also necessary for him, in the film's final moments, to terminate himself. Like certain other saviours, he must die (or, at least, be permanently shut down) in order for humanity to live. By becoming both mortal and self-sacrificing, Arnie completes the transition from amoral automaton to heroic humanoid.

Terminator 2's baddie, by contrast, is a miracle of 1990s' special effects and inhuman efficiency. Formed from liquid metal, the T-1000 manifests its malignancy through mimicry, becoming a simulacrum of successive authority and parental figures: a police officer, John Connor's stepmother, Sarah Connor herself. As merciless and implacable as Spenser's Talus, its every action is bent towards not Justice but the termination of John Connor – because he will, as an adult, lead the human resistance against the AI that created it (and which has thus sent the T-1000 back in time to eliminate him as a child).

Like all good action movies, *Terminator 2* focuses on a handful of characters committing spectacular acts of violence, but its existential stakes preserve an impressive degree of uncertainty.[19] In a series of apocalyptic visions, Sarah Connor foresees the 'Judgment Day' soon to be brought about by the aforementioned AI, Skynet, which we learn is going to slaughter most of humanity with nuclear weapons two years after the film's main action.

It's a sufficiently influential version of the Singularity myth that 'building Skynet' has become shorthand for creating a malevolent supercomputer. But it's also a future that the movie's heroics don't so much eliminate as downgrade from inevitable to possible. 'The future's not set,' Sarah Connor repeats at several key moments. 'There's no fate but what we make for ourselves.' These aren't phrases that guarantee either apocalypse or salvation. Rather, they proffer uncertainty as our greatest prize: a blend of freedom and responsibility defined by the belief that it isn't yet too late.

By the end of the film, we find ourselves back at square one: unaccompanied by a superhuman saviour or destroyer, condemned to be free. It is a fine place for storytelling to end and life to begin.[20]

How to think about the future

To think magically is to believe literally in things that cannot literally be true: to succumb to the delusion that our creations can gift us both purpose and salvation. This kind of magic explicitly rejects reality in favour of seductive fantasy. There's another way of thinking about magic in the context of technology, however, that owes its most famous formulation to the author Arthur C. Clarke. It's sometimes known as Clarke's third law and goes like this: 'any sufficiently advanced technology is indistinguishable from magic.'[21]

Clarke set out this so-called 'law' in a later footnote to his 1962 essay 'Hazards of Prophecy: The Failure of Imagination', the subtitle of which signals his intentions. Magic, here, denotes not so much wishful belief as the *refusal* to believe in some remarkable future possibility.

> Suppose you went to any scientist up to the late nineteenth century and told him: 'Here are two pieces of a substance called uranium 235. If you hold them apart, nothing will happen. But if you bring them together suddenly, you will liberate as much energy as you could obtain from burning ten thousand tons of coal.' No matter how farsighted and imaginative he might be, your pre-twentieth century scientist would have said: 'What utter nonsense! That's magic, not science. Such things can't happen in the real world.'[22]

Building on the theme of nuclear energy, Clarke considers the case of the great scientist Ernest Rutherford. Through the first decades of the twentieth century, Rutherford helped to lay the foundations of modern nuclear physics – including, between 1908 and 1913, overseeing experiments by Hans Geiger and Ernest Marsden[23] that were the first to find direct evidence of atomic nuclei, as Rutherford himself had predicted in a seminal 1911 paper.[24] In 1933, however, Rutherford commented during a public lecture that the chance of finding 'sources of power in atomic transmutations' was 'the merest moonshine',[25] a view he maintained until his death in 1937. The vast energies locked inside atomic nuclei, he suggested, would never be harnessed as a power source.

It was an assertion that was soon decisively disproved, with the discovery of uranium fission in 1939 followed by the development and unleashing of atomic bombs and, in due course, the creation of nuclear reactors. Rutherford had, in Clarke's opinion, allowed his intimate understanding of present realities to blind him to future possibilities: 'too great a burden of knowledge can clog the wheels of imagination.'[26]

Clarke is surely correct that the future must, sooner or later, outstrip all that any present can hope or know. Yet perhaps the most important lesson his essay has to offer is that while *prediction* is only possible under certain conditions, *imagination* may yet equip us to confront time's unknowable, indefinite unfolding. For while any literal attempt at prophecy will sooner or later founder, credulity or wilful blindness aren't the only alternatives. Rather, via the sufficiently careful application of imagination, we can find our way towards enduring questions, tensions and possibilities.

Consider what it means successfully to practise the art and science of prediction. As the political scientist and author Philip Tetlock has argued in his work on 'superforecasters' (individuals who are unusually skilled at making predictions), accurate forecasts tend to entail an excellent grasp of the big picture, an eagerness to learn and adapt in the light of new evidence and an open-minded interest in a diversity of perspectives. Above all, however, the possibility of prediction relies upon how amenable a field is to prediction in the first place. The more regular and law-like its features, the better. The more chaotic and uncertain it is, the more broadly predictions have to be couched – and the shorter the horizons to which they apply.[27]

In this sense, Rutherford's 'moonshine' comment was not so much misguided as miscategorized. It was made in the context of efforts to trigger nuclear reactions via a proton beam in the first particle accelerator in 1932: a machine which did indeed require vastly more energy to operate than it could ever produce.[28] Within the context of that particular technology, Rutherford's verdict was correct. The problem wasn't so much his expert assessment of the evidence as his underestimation of the speed at which this frame of reference would be replaced by a new paradigm. What he should have done (an easy enough phrase to use in retrospect) was look to the longer timescales within which scientific revolutions play out – and the ways in which paradigm shifts are *themselves* a prediction we can make with confidence.[29]

It's relatively common today to consider the history of science and technology in terms of such revolutions – an approach most often associated with Thomas Kuhn's 1962 book *The Structure of Scientific Revolutions*, and with his observation that the development of scientific understanding entails not so much the steady accretion of knowledge as radical discontinuities between different eras. Kuhn, a physicist by training, drew his inspiration from a version of the very problem Clarke identified: that countless brilliant thinkers from previous eras defended claims that have come to seem not only wrong but also demonstrably ridiculous. As Kuhn put it in a later reflection upon the inspirations for his book:

> I found it bothersome . . . [that] Aristotle appeared not only ignorant of mechanics, but a dreadfully bad physical scientist as well. About motion, in particular, his writings seemed to me full of egregious errors, both of logic and of observation.[30]

How could Rutherford have failed to grasp the implications of his own discoveries? How could Aristotle have misunderstood something as simple as motion? In Aristotle's case, Kuhn suggests, the answer is that the very concept of 'motion' didn't and couldn't mean the same thing to him as to a twentieth-century scientist.

Two and a half thousand years ago, any deep consideration of the phenomenology of movement necessarily encompassed living as well as inanimate systems. It entailed discussions of desire, nature and inclination: it was as much a matter of natural history and metaphysical speculation as mechanical description. Indeed, the very notion of a mechanical description in the post-Newtonian sense would have been nonsensical to Aristotle. As we all do, he existed within the narratives and assumptions of his age.

So long as we're aware of this, there remains a great deal we can learn from Aristotle – but only if we resist the category error of misreading our *own* assumptions as a final and literal form of truth. Indeed, it was the revelation that Aristotle was 'a very good physicist indeed, but of a sort I'd never dreamed possible'[31] that Kuhn goes on to describe as his key insight. As the author and physicist Carlo Rovelli has also argued, there's something fundamentally ahistorical about the claim that it was the refutation of Aristotle's errors that drove scientific progress two millennia later. Far more significant are the continuities, dialogues and common imaginative ground within humanity's ongoing efforts to understand the world.

> Do objects of different weight fall at the same speed? At school we are told that, by letting balls drop from the tower of Pisa, Galileo Galilei had demonstrated that the correct answer is yes. For the preceding two millennia, on the other hand everyone had been blinded to the fact by the dogma of Aristotle, according to which the heavier the object, the faster it falls . . . It's a good story, but there's a problem with it. Try dropping a glass marble and a paper cup from a balcony. Contrary to what this beautiful story says, it is not at all true that they hit the ground at the same time: the heavier marble falls much faster, just as Aristotle says . . . Aristotle did not write that things would fall at different speeds if we took out all the air. He wrote that different things fall at different speeds in our world, where there is air. He was not wrong.[32]

Ultimately, Galileo was able to build upon Aristotle's insights not because he treated them as regrettable delusions, but because he understood them subtly and intimately enough to tease out the possibilities latent within them: to pursue details and anomalies in the light of new knowledge. The same is true of Newton's readings of Galileo; of Einstein's readings of Newton; and of countless other unsung re-readings of older theories and speculations.

In this sense, imagination's most important task has little to do with prediction and everything to do with warnings *against* unwarranted certainty. It is there to remind us that the past is all we have when it comes to grasping possible futures, and that our present will soon enough become someone else's history. Its task is to help us entertain profoundly different perspectives to our own – and to conjure, within their variety, stories that speak to the grandest timescales of all.

Coming to terms with our place in the universe

What form might these stories take? As Kuhn suggests, they need to encompass a history of rivalrous and evolving world-views: shifting perspectives upon time, technology and history that mark our present assumptions as both better informed than any that have come before, yet equally transient in their particulars.

In his 1917 *General Introduction to Psychoanalysis*, first published in English in 1920, Sigmund Freud offered one of the last century's most influential accounts of the evolving relationship between scientific discovery and human self-understanding. For Freud, the story of our increasing knowledge of the universe is above all characterized by a radical *decentring* of humanity; by the diminution of our significance brought by new knowledge.

> Humanity, in the course of time, has had to endure from the hands of science two great outrages against its naive self-love. The first was when humanity discovered that our earth was not the centre of the universe but only a tiny speck in a world system hardly conceivable in its magnitude. This is associated in our minds with the name Nicolaus Copernicus, although Alexandrian science had

taught much the same thing. The second occurred when biological research robbed man of his apparent superiority under special creation, and rebuked him with his descent from the animal kingdom and his ineradicable animal nature.[33]

The Copernican and Darwinian revolutions, Freud immodestly suggested, were being followed by a third revolution in the form of his own psychological research – and its revelation that the conscious mind 'is not even master in its own home, but is dependent upon the most scanty information concerning all that goes on unconsciously in its psychic life'.[34] We cannot, in other words, truly know even ourselves.

In his 2014 book *The Fourth Revolution*, the philosopher Luciano Floridi builds on Freud's account to suggest that the information age heralds a fourth such transformation; one characterized by a further humbling. Here is Floridi's synopsis of what this revolution entails in the context of those that came before:

> today we acknowledge that we are not immobile at the centre of the universe (Copernican revolution), that we are not unnaturally separate and diverse from the rest of the animal kingdom (Darwinian revolution), and that we are far from being Cartesian minds entirely transparent to ourselves (Freudian or neuroscientific revolution) . . . [now] we are slowly accepting the post-Turing idea that . . . we are informational organisms (inforgs), mutually connected and embedded in an informational environment (the infosphere), which we share with other informational agents, both natural and artificial, that also process information logically and autonomously . . .[35]

Floridi's fourth revolution suggests that we can no longer presume to be our planet's sole site of analysis and reason. This is because our computational creations are approaching or exceeding our capabilities in areas long believed to be uniquely human: calculation, deduction, recall; language processing, pattern recognition; the modelling and prediction of the world.

What follows from this? Among other things, Floridi suggests that something 'intrinsically informational' about human nature is being laid bare – and that we must now seek to understand ourselves as 'informational organisms', our minds and capabilities not only enhanced by technology but also interwoven with it. To a degree that even the extended mind hypothesis only hints at, our decisions and freedoms do not belong to us alone. They are facets of the information environments we inhabit: of the decision-making agents that populate these realms in their swelling millions; of the layers of data woven between us and the world.

Floridi's updating of Freud is a resonant way of putting the present in its place. And it emphasizes that although world-views come and go, this does not mean they are all created equal. There is, rather, a fundamental divide between accounts of our place in the universe that *deny* inconvenient information and accounts that seek to *accommodate* and *account for* it: between the magical and the mythic, the delusory and the self-knowing. Indeed, on a species-wide scale, the fundamental question we all face echoes one constantly confronted by each individual consciousness: what does it mean to update the 'controlled hallucination' of self-perception in the light of new knowledge?

If you are still feeling mythologically inclined, the sequence of shocks underpinning recent history looks like what the Germans call a Bildungsroman – a tale of maturation from childhood to adulthood. The notion that we might 'grow up' as a species is one this book has invoked several times, alongside the humility it demands: the acknowledgement, in Freud's stark phrase, of multiple 'outrages' upon our 'naive self-love'. It is no longer reasonable for us to regard our planet as the centre of the universe, our species as uniquely created to rule over nature, our minds as sublimely lucid and self-knowing, or our intellectual capacities as beyond replication. These are the facts we must integrate into the telling and retelling of our species' story.

Despite Freud's protestations, however, there's no rule that humility should be synonymous with humiliation – nor a naturalistic understanding of our universe incompatible with grander purposes. On the contrary, the miraculous extent of our mutual dependency and continuity with other life has been laid bare. Like Didion wrestling with the magical mindset that left her 'thinking as small children think', we must find a way to put our loves and longings where they belong: in a place that acknowledges their supreme value without pretending this grants them supernatural force.

Neither nature nor existence has a particular plan for people. Our creations grow faster than we do, and may reach further. Yet we are all the more remarkable for this – if we can learn to let go of the insistence that it all comes down to either a battle or a love affair. We are tool-making, technological creatures; hyper-social entities; storytellers by nature, protean, grappling with a dream of what we believe ourselves to be. Reduced to the level of individual users tapping and clicking at screens, we seem all too fragile: antiquated devices ripe for displacement. But together, weaving a semi-autonomous web of information across our world, we are something else: miracles of life and comprehension; uniquely capable, burdened and brilliant.

CHAPTER 9

Trickery and intellect:
The anthropomorphic delusion

In 1770, the inventor Wolfgang von Kempelen displayed a mechanical marvel to the Viennese court. Watched by the Archduchess Maria Theresa and her entourage, he opened the doors of a wooden cabinet four feet long, three feet high and just over two feet deep, illuminating its interior by candlelight to display glistening cogs and gears. Seated at the cabinet was a life-size model of a man in Turkish dress – a turban and fur-trimmed robe. In front of the Turk, on top of the cabinet, was a chessboard.

Kempelen closed his cabinet and asked for a volunteer to play a game of chess against the Turk. It was an astonishing request. Finely crafted automata had been entertaining the nobility and paying public for decades, but the idea that one might undertake an intellectual task such as chess was unprecedented: something for the realm of magic rather than engineering. This was precisely the point. Six months previously, Kempelen had told the Archduchess that he was utterly unimpressed by a magic show performed for the court by a French conjurer, and could create a finer spectacle himself. The Turk was the result.

Count Ludwig von Cobenzl, the first volunteer, approached the table and received his instructions: the machine would play white and go first; he must ensure he placed his pieces on the centre of each square. The count agreed. Kempelen produced a key, wound up his clockwork champion, and with a grinding of gears the match began. To its audience's astonishment, the machine did indeed play, reaching out with a gloved hand to move piece after piece. It even nodded its head to indicate check. Within an hour, the count had been defeated, as were almost all the Turk's opponents during its first years of growing renown in Vienna.

In 1781, just over a decade later, Maria Theresa's son, Archduke Joseph II, asked Kempelen to present his creation to its most illustrious audience yet: Catherine the Great's son and heir, Grand Duke Paul of Russia. Kempelen was reluctant. A serious scientist and engineer, he had become embarrassed by the spell his creation seemed to cast and had taken to claiming it no longer worked. At Joseph's command, however, the Turk was restored in time for the Grand Duke's arrival. It proved a sensation. Indeed, Joseph was so delighted that he told Kempelen to take the Turk on a tour of Europe, a request Kempelen (once again) had little choice but to obey.

In 1783, adapted for travel and packed into boxes, the Turk and its creator set off. Having entertained the French court at the Palace of Versailles, the Turk was put on public display in Paris, a city renowned for the skill of its chess players. Once again, it delighted and astonished audiences – and won almost all of its matches. Its status, meanwhile, was hotly debated. Could the workings of the human mind really be replicated by clockwork? Theories ranged from pre-programmed positions subtly selected by Kempelen to a child or dwarf hidden inside the mechanism, but a definitive answer remained elusive. Over the following decades, the Turk travelled through London and Germany, inviting fervent speculation wherever it went. Among its losing opponents were Benjamin Franklin, visiting Paris in 1783, and – under its second owner after Kempelen's death – the Emperor Napoleon in 1809. Napoleon allegedly tested the machine with illegal moves, only to see the Turk sweep the pieces off the board in apparent protest.

The Turk was, of course, a fraud: a magic trick masquerading as a mechanism. Behind the cogs and gears lay a secret compartment, from within which a lithe grandmaster could follow the game via magnets attached to the underside of the board, moving the Turk's arm through a system of levers. In his 2002 book *The Mechanical Turk*, the British author Tom Standage tells the automaton's story in captivating detail, noting that even the publication in 1821 of an account accurately explaining why the Turk *must* contain a hidden operator scarcely diminished its power. The image of man and machine locked in combat across the chessboard was simply too perfect, and too perfectly matched to the unease growing in the late eighteenth century around technology's usurpation of human terrain.

Kempelen's desire to make not only a machine but also a kind of magic was also no accident. In 1971, a prop builder for professional magicians, John Gaughan, set out to create a full reconstruction of the Turk. He refined his creation for eighteen years until, in November 1989, he finally presented its first public performance at a Los Angeles conference on the history of magic. Even though everyone present knew it was a trick, the result remained remarkably – even eerily – compelling. As Standage explains, the sheer impact of the performance emphasized the degree to which the Turk was as much a psychological as a technological marvel; and how its success relied upon subtle framing effects, showmanship and misdirection.

> Gaughan's view is that the Turk is primarily an example of a magician's rather than an engineer's ingenuity . . . For starters, he points out, [its final owner] used to display the Turk at the end of a show that consisted of several genuine automata, to get the audience into the right frame of mind, so that they assumed that there was no

limit to what mechanical contraptions could do . . . Another important aspect of the illusion is the loud noise made when winding up the automaton . . . This noise, and the physical effort required to turn the handle, would have strengthened the spectators' belief that the clockwork machinery shown to them before the game began was responsible for determining the Turk's moves . . .[1]

Kempelen had grasped two intriguing points about human attitudes towards technology: that what we *think* a machine is doing need not have anything to do with what is *actually* going on; and that we're only too keen to attribute human-like autonomy to machines, in the process forgetting how much they depend upon their makers and maintainers. As Standage puts it:

the illusion it is genuinely a pure machine is extraordinarily compelling, even to those who know how it works. Something about the Turk seems to evoke a fundamental human desire to be fooled.[2]

Denials and enchantments

There's nothing inherently wrong with deception, so long as you have agreed that this is what you want. Like the suspension of disbelief entailed by art, making magic necessarily begins with the desire to be fooled, for without this there could be neither impetus nor interest. To watch stage magic is to witness the seemingly impossible, to wonder how it might have been achieved, then to delight in the fact that you cannot answer your own question.

The corollary to this is that magic must be crafted with great skill for the trick to work. The last thing a professional magician can afford is to think magically, in the sense of relying on hope. Denials and delusions are for those at the receiving rather than the engineering end of enchantment. The resulting asymmetry between audience and performer is as stark as it comes. Thousands of hours of practice lie behind the best performances, along with props, methods and gimmicks shrouded in secrecy. Behind the scenes, the mechanics of a laboriously rehearsed show unfold like clockwork. Every viewing angle, potential error and giveaway must be anticipated; every gesture and distraction tailored for maximum impact. From the audience's perspective, meanwhile, the result is a steady stream of marvels. The more effortless and incredible it seems, the more intensive the labour that has gone into making it that way.

This is as true of twenty-first-century technologies as eighteenth-century magic shows. As Standage notes, the last few centuries of technological modernity have played out 'at the intersection between science, commerce, and entertainment.'[3] Throughout the emergence of modern industry and communications across the eighteenth and nineteenth centuries, electrification's cutting edge was as much a matter for carnivals and bravura demonstrations as sober research, with its wonders deployed to equally sensational effect by charlatans, researchers and salesmen. Here's the author and professor of communications Carolyn Marvin describing some of these techniques.

> The 'finale' of an electrical lecture by Edison Company representatives in Boston in 1887 was a spiritualistic seance. 'Bells rang, drums beat, noises natural and unnatural were heard, a cabinet revolved and flashed fire, and a row of departed skulls came into view, and varied colored lights flashed from their eyes.' So impressive were these effects that a second lecture had to be repeated, for a standing-room-only audience. In 1894 the Sunday World attached this caption to a picture accompanying an article by Arthur Brisbane about a lecture by Nikola Tesla: 'Showing the Inventor in the Effulgent Glory of Myriad Tongues of Electric Flame After He Has Saturated Himself with Electricity.'[4]

Pioneers like Nikola Tesla were at once geniuses and showmen, inventors and magicians. It's not for nothing that the twenty-first-century car company named after Tesla embodies a heady mix of hype and innovation. Indeed, the notion that we've moved beyond 'debased' shows and tricks shouldn't survive a moment's consideration of twenty-first-century advertising, entertainment and interfaces. By tapping into enchantment – by speaking to our tendency systematically to misread machines' mechanisms and capabilities – engineers and corporations increase the allure of their creations while escaping critical scrutiny. And the sophistication with which this is done makes yesterday's 'fake mysteries' look trivial by comparison.

Arthur C. Clarke's claim that any sufficiently advanced technology is indistinguishable from magic takes on a new light in this context. After all, 'real' magic is always and only a mix of skill, gadgetry and obfuscation – plus a human audience who are unable to explain what they've seen. The asymmetry between magician and audience (or between engineers and users) is the key. Once the secret is known, you're left with nothing more than the effect of one person's ingenuity upon another.

It's this shift in perspective that turns yesterday's enigmas into today's self-evident shams. Kempelen pretended that his Turk could play chess, and this pretence was powerful both because of his stagecraft and because automatons' potentials remained imperfectly understood in the late 1700s. By the late nineteenth century, the limitations of clockwork had become clearer. Multiple experts had confidently determined that the Turk was a parlour trick – and that the very prospect of such a machine playing chess to a high standard was impossible. Standage cites the automaton-maker Charles Godfrey Gümpel who, in 1879, produced a proof showing that any attempt physically to represent even a fraction of chess's possible configurations would entail billions of people working for millions of lifetimes. Clearly, a mechanism able to master so complex a conundrum couldn't exist. The only thing capable of challenging a human mind was another human mind.[5]

Is it reasonable, here, to raise a retrospective smile at the limitations of nineteenth-century expertise? Armed with over a century of hindsight, it may seem obvious to us that mechanical automatons cannot play grandmaster-level chess while electronic computers can. Yet the deeper point is the same as it has always been. Attempting to predict technology's long-term path is a losing game; but enduring insights can be found if we resist false confidence and look to human continuities. Why was the Turk so convincing? Why were the true potentials of computation misunderstood? Because the paired tendencies that Kempelen exploited remain only too significant: our eager projection of human attributes onto non-human entities; and our related inability to grasp the inhuman means through which human achievements might eventually be replicated – or exceeded.

Seductive technologies and the Turing test

One of the most intriguing places the above confusions come together is a seminal twentieth-century thought experiment: the Turing test. 'I propose to consider the question, "Can machines think?"' wrote the pioneering computer scientist Alan Turing in the first line of his 1950 paper 'Computing Machinery and Intelligence'. The trouble with such a question, Turing observed, was that answering it was likely to involve splitting hairs over the meaning of words like 'machine' and 'think'. Thus, he continued:

> I shall replace the question by another, which is closely related to it and is expressed in relatively unambiguous words. The new form

of the problem can be described in terms of a game which we call the 'imitation game'.[6]

In its original form, Turing's imitation game entailed a conversation between a human tester and two hidden parties, each communicating with the tester via typed messages. One hidden party would be human, the other a machine. If a machine could communicate such that the human tester could not tell which of their interlocutors was which, the machine would have triumphed. To be precise, Turing first suggested that the hidden parties would be a man and a woman, then that a machine could be swapped for one of these. This was presumably meant to make his proposal easier to grasp by likening it to a parlour game – an analogy which, in an age when playing with computers has become more common than playing in parlours, hasn't stood the test of time.

Such a game was a useful test of intelligence, Turing suggested, because it abandoned subjective definitions in place of a challenge that could be passed or failed, not to mention endlessly restaged. By excluding semantic ambiguities in favour of a staged performance, it made the ineffable conceivable. But it also – as Kempelen had intuited – brought with it the problems inherent to any staged contest.

As soon as you've turned something into a game, any legitimate move that brings you closer to victory is fair play. If victory is defined as successful deception, the techniques likely to achieve this will owe as much to stagecraft as to technology. And if only one of the parties involved can be 'fooled', this introduces a fundamental asymmetry. As Turing noted, machines can be designed to simulate human limitations, but humans cannot be redesigned to out-calculate machines:

The game may perhaps be criticised on the ground that the odds are weighted too heavily against the machine. If the man were to try and pretend to be the machine he would clearly make a very poor showing. He would be given away at once by slowness and inaccuracy in arithmetic.[7]

Turing's game was, for the best part of three decades, played for real during the annual Loebner Prize, which between 1990 and 2019 offered an award to the AI chatbot able to convince the most judges that it might be human. To the frustration of many specialists, some of the most successful early chatbots used tricks based on stock responses and emotional impact rather than complex code: bogus bios and pre-programmed typing errors; hesitations, colloquialisms and insults.[8] One impressive performer, winning second place in 2005 and 2008, was a chatbot pretending to be a thirteen-year-old Ukrainian boy called Eugene Goostman. By playing the role of a teenager typing in a second language, it managed to fool a surprising number of humans even when its responses were non-sequiturs. As its creators wrote in a 2008 essay:

> 'The imitation game' is exciting, amusing and highly intelligent – but it is nothing but a game. Do not expect that passing it means anything more than that some bot was luckier than the rest . . . When making a bot, you don't write a program, you write a novel. You think up a life for your character from scratch – starting with his (or her) childhood and leading up to the current moment, endowing him with his personal unique features – opinions, thoughts, fears, quirks.[9]

Given the history of stage magic, these manipulations were predictable enough. Indeed, versions of them remain central to the hacking skillset known as 'social engineering', which relies on fooling a system's human users in order to extract information. More unexpected, however, is the effectiveness of a strategy targeting an orthogonal aspect of the narrative imagination: the desire not to be told a story, but to have someone listen to your own.

This was the insight behind one of the first and most famous of all chatbots, ELIZA, created in the mid-1960s at the MIT Artificial Intelligence Laboratory by the computer scientist Joseph Weizenbaum. ELIZA was designed to interact according to a range of scripts, the most famous of which simulated a form of psychotherapy by reflecting back key words from users' inputs. Here is a partial transcript of one of its conversations, as reported in Weizenbaum's 1976 book *Computer Power and Human Reason*, with the labels Human and Eliza inserted to identify each party.

understanding to it – and continued to insist on this even when they were shown the extent and limitations of its code. Third, its successes as a conversational agent led some to claim that computers could in principle 'solve' natural language, despite Weizenbaum's insistence on the very opposite: that no such general solution was possible, let alone prophesied by his simple rules-based program.

In the half-century since ELIZA's creation, computers' capabilities have advanced further than many dreamed possible. Yet it remains striking both how compelling the illusion of understanding conjured by ELIZA's interactions remains, and how relevant Weizenbaum's concerns are to the latest generation of AIs. To take them in reverse order, machines are now able to demonstrate a remarkable command of language; but the ways in which they achieve this are utterly unlike human understanding. Humans' emotional involvement with machines is more significant than ever; but machines' emotional involvement with humans remains precisely zero. And the automation of everything from care and therapy to customer support has proved eminently feasible; but it's also profoundly controversial, not least because what's on offer is not so much about machine intelligence as about the management of human emotion and expectation.

Weizenbaum chose the name ELIZA because this was what George Bernard Shaw called the heroine in his 1913 play *Pygmalion*, a re-telling of the eponymous myth. Unlike Ovid's Galatea, however, Shaw's Eliza is always and only a flesh-and-blood woman: a 'common flower girl' whom the play's male protagonist, Professor Henry Higgins, claims he can remould into a genteel lady. Shaw's class-conscious reframing of the story struck a chord with twentieth-century audiences. When it comes to technology, however, what's most intriguing about his updated version is its replacement of a literal transformation with a project predicated on manipulation, deception and self-flattery. Higgins isn't a sculptor blessed with divine talent: he's an arrogant misogynist determined to demonstrate his own cleverness. Similarly, one of the great lessons of programs like ELIZA has turned out to be not that machines can be transmuted into genuinely empathetic conversationalists, but that the combination of wishfulness and wonder brought by humans to such interactions can be staggeringly effective at masking what's *really* going on.

Consider modern 'virtual assistants' like Amazon's Alexa, Apple's Siri and Google's Assistant. Controlled largely through voice commands, these domesticated AIs don't (yet) try to fool you into thinking you're dealing with either a human or a truly intelligent machine. Instead, the focus is on creating a user experience that combines access to proprietary services with as high a level of comfort, ease and delight as possible. A significant part of this is a profusion of witty canned responses (sometimes called 'Easter Eggs') designed expressly to engage the emotions rather than the intellect.

Ask Amazon's digital personal assistant Alexa whether 'she' can pass the Turing test and you'll get the reply, 'I don't need to pass that. I am not pretending to be human.' Ask Apple's Siri if 'she' believes in God and you'll be told, 'I would ask that you address your spiritual questions to someone more qualified to comment. Ideally, a human.' Ask Google's Assistant whether it's actually Skynet, and it'll reply, 'No way. I like people. Skynet hates people. I rest my case.' These and countless other responses have been prescripted to amuse and to disarm. They're meant to fool us into perceiving not intelligence but innocence: products too charming and too useful to provoke any deeper anxiety.[11]

Like many parents, I listen to my young children chatting to virtual assistants with a mixture of amusement and unease. My son and daughter simultaneously treat them as toys and confidants; as boxes of tricks and oracles. Like a miniature version of Arnie in

Terminator 2, they are simply *there*: listening with infinite patience; ready endlessly to crack jokes and spit out facts; surreptitiously hungry to harvest data about domestic life. In terms of user experience – from the perspective, that is, of their audience rather than their engineers – they are more like pets than people, and all the more enchanting for it. To quote the author and psychologist Sherry Turkle:

> Technology is seductive when what it offers meets our human vulnerabilities. And it turns out we are very vulnerable indeed.[12]

Today's machines put on a magnificently disarming show. And the same is true of artificial entities conceived for very different purposes: bots spreading disinformation on social media; bogus personalities designed to debase discourse to a level where machines and humans *are* indistinguishable; the tirelessly accommodating glamour of AI-generated influencers.

This is the twenty-first-century art of user experience in all its glory: a hall of mirrors within which *who* and *what*'s real is, increasingly, impossible to determine. To borrow a phrase from the philosopher Charles Taylor, ours is an age of *re-enchantment*: its infrastructure suffused with secret purposes, its surfaces concealing fractal depths.[13] We are surrounded by exhortations towards magical thinking: by showmanship and social engineering. And this in turn obliges us to look deeper and harder in search of answers: to the racked servers and petabytes of data within which information technology's magic is truly made.

Losing to computers at chess

To return to the scenario with which this chapter began, what does it mean to play a modern computer at a game like chess? Above all, it means losing. In 1997, humanity's greatest chess champion, Garry Kasparov, was beaten before the eyes of the watching world by IBM's Deep Blue. In 2016, Deep Mind's AlphaGo did the same for Go champion Lee Seedol, besting humanity at a game orders of magnitude more complex than chess. In early 2017, an AI called Libratus, developed at Carnegie Mellon University, vanquished the world's best players at no-limit Texas Hold 'Em, a game of bluff and imperfect information that some had hoped would remain dominated by humans. It was also in 2017 that Deep Mind's AlphaZero mastered Chess, Go and Shogi (a Japanese game closely related to chess) purely by playing against itself; and in 2019 that Deep Mind's AlphaStar first beat a top human professional at the real-time strategy video game StarCraft II.[14] By 2022, Meta's Cicero AI was able to play the strategic game Diplomacy at a high human level – including negotiating alliances and engaging in open-ended chats with human players.[15] By the time you read these words, further citadels of human expertise will surely have fallen.

This progression points both to the startling nature of recent advances in AI and to a fundamental divide between people and machines. Much like athletes pushing against the boundaries of biology, the increments of human improvement have hard limits. We advance towards a certain threshold in slowing steps. Across rapid generations of software and hardware, meanwhile, machines advance faster and faster. Since 1997, the world's best human chess players have got perhaps a little better, helped by computer analyses. Meanwhile, the speed at which Deep Blue calculated (around eleven Gigaflops: each Gigaflop represents one billion floating point operations per second) has fallen several orders of magnitude behind even mobile phones (Apple's 2023 iPhone 15 Pro Max manages over two thousand Gigaflops) let alone supercomputers (Frontier, the world's fastest computer as of early 2023, is capable of over 1.6 Exaflops, each representing a billion Gigaflops). The Deep Blue of 1997 would stand as much chance against an app running on today's smartphones as a two-year-old would against Kasparov.[16]

Perhaps it's misleading to present this as surprising. As the computer scientist Stuart Russell points out in his 2019 book *Human Compatible*, the above results are entirely in line with a steady rate of increase in computers' game-playing performance since the mid-1960s. What makes it *seem* incredible is the fact that this kind of progression defies human intuition. As Russell puts it:

> For AI researchers . . . the real breakthroughs happened thirty or forty years before Deep Blue burst into the public's consciousness. Similarly, deep convolutional networks [which underlie many recent AI 'breakthroughs'] existed, with all the mathematics fully worked out, more than twenty years before they began to create headlines.[17]

How should we think about such change? To continue the chess analogy, the futurist Ray Kurzweil in his 1999 book *The Age of Spiritual Machines* uses the phrase 'the second half of the chessboard' to help readers conceptualize it. This refers to a mathematical parable in which a scholar is told by a king that he can name any price as his reward for performing a great service. 'What I wish for,' the scholar replies, 'is that you should place one grain of wheat upon the first square of a chessboard, two upon the second, four upon the third, eight upon the fourth, and so on, until the chessboard is covered.' The king protests that this is too small a prize, but the scholar demurs. By the end of the first row of eight squares, he has 255 grains of wheat. By the time the first half of the chessboard is covered, he has 4,294,967,295 – around 280 tonnes. After this, the first square on the second half of the chessboard will contain as much wheat as the entire first half, and so on, until the amount exceeds all the wheat that has ever existed (the final total is around 1.2 trillion tonnes).

This accelerating departure from history and conceivability is the point of the fable. The first half of the chessboard is compatible with everyday experience and intuitions. If you keep going for long enough, however, an exponential increase will demolish every old frame of reference. As Kurzweil, writing on the eve of the twenty-first century, put it:

> In many areas – the ability to write a book on computers, for example – computers are still pathetic. But as computers continue to gain in capacity at an exponential rate, we will have the same experience in these other areas that Kasparov had in chess. Over the next several decades, machine competence will rival – and ultimately surpass – any particular human skill one cares to cite, including our marvellous ability to place our ideas in a broad diversity of contexts.[18]

Can we hope for anything other than obsolescence in the face of this acceleration? This book's first chapter posed a version of this question, as well as suggesting that Kurzweil's account of a point of no return – the so-called technological Singularity – is at once profoundly significant and, at least in its literal form, misguided. Time and technology may prove otherwise. But it's important to remember that the very use of games as emblems of humanity's relationship with machines brings with it a host of assumptions and associations, not least the atavistic image of a lone human locked in combat with an automaton. To compete within a bounded, predictable arena is indeed to engineer the circumstances of our own defeat. But it is also to ignore those differences between minds and machines – not to mention the possibilities of collaboration between them – that, in the long term, matter most of all.

How AIs relate to the world

While the Singularity may not be coming, the vision of technological apotheosis it embodies is undoubtedly worth taking seriously, not least because — at least within the bounds of game-worlds like chess and Go — the Singularity has indeed come and gone. Never again will the world's greatest player be an unaided human. And, just as Kurzweil predicted, this is becoming true of more and more domains.

When, for example, I sit at my desk and interact with the Large Language Model known as ChatGPT, which took the internet by storm upon its public release in November 2022, it can be impossible to distinguish many of its outputs from something a well-informed human might write. Here's just one example. What follows is my initial prompt, followed by the system's unmodified output.

> TOM: Describe the chess match between Deep Blue and Garry Kasparov.
>
> ChatGPT: The chess match between Deep Blue and Garry Kasparov was a historic six-game chess match that took place in May 1997 in New York City. The match was a rematch of the 1996 match between Kasparov and Deep Blue, in which Kasparov won 4–2. Deep Blue was a computer program created by IBM that was designed to play chess at a high level . . . The 1997 match was highly anticipated, as it pitted the world's best human chess player against a machine that had been designed specifically to beat him. The match began on May 3, 1997, and was played over six games.[19]

A Large Language Model, or LLM, is a subtype of Artificial Intelligence algorithm that has been trained on massive quantities of text, and that can fluently summarize and generate new text in response to prompts. The techniques underpinning LLMs – such as deep learning and neural networks – have been studied for decades; but a significant recent breakthrough was the introduction of a deep learning architecture based on so-called 'transformers' in 2017, which use 'attention' mechanisms to train algorithms on linguistic data faster and more flexibly than previous approaches. GPT stands for Generative Pretrained Transformer, and ChatGPT's capabilities (not to mention those of its successor, GPT4) far exceed those of any system from even a few years previously.[20] A crucial question that comes with this power, however, is how closely its outputs are – or aren't – connected to reality. In the above interaction, all of the factual details ChatGPT supplied were correct. By contrast, here is my follow-up request.

TOM: List some interesting academic analyses of the match.

ChatGPT: 'Man vs. Machine: An Historical Overview of Deep Blue vs. Garry Kasparov' by Murray Campbell, A. Joseph Hoane Jr., and Feng-hsiung Hsu: This paper, published in the journal Artificial Intelligence, provides an in-depth technical overview of the development of the Deep Blue chess computer, and the strategies it used to defeat Kasparov.

'The Brains Behind the Game: Evaluating the Role of Chess Knowledge in a Chess-Playing Computer' by James R. Slagle: This paper, published in the Journal of Experimental Psychology, examines the role of chess knowledge in the performance of Deep Blue, and argues that Deep Blue's success was due to a combination of advanced algorithms and human-designed heuristics.

'When the World Chess Champion Played the Chess-Playing Computer' by David Levy . . . [21]

This, too, is a coherent and expert-sounding response. In total, ChatGPT offered five academic references in response to my query, and it's difficult at a glance to distinguish its recommendations from the verdict of a well-informed human. Yet not one of them is an actual research paper. So what's going on?

The short answer is that systems like ChatGPT don't exist in the same world as us. Humans live, perceive and strive from a particular perspective, anchored to the world by millennia of evolutionary struggle. Our creations, however, are the children not of biology but information: utterly dependent upon the inputs we define. Hence the mesmerizing stream of so-called 'hallucinations' created by the latest generation of LLMs – and the larger category of algorithm to which they belong, Generative AIs, which can plausibly fabricate images, faces, words, video and audio, and have begun doing so on an industrial scale.

ChatGPT's answer to my second query is a mélange of possibilities and precedents. Most of the authors it lists have published papers in relevant fields, while several actual papers closely resemble the ones it describes. Yet its response exists slantwise to our world. It's not so much a series of factual claims as a riff upon billions of data points proximate to my input.

Despite the vividness of the term 'hallucination', it's also worth remembering that what's occurring within an AI is nothing like human consciousness. A system like ChatGPT is designed to generate statistical inferences based upon the data it has ingested. In terms of its internal logic, it will always do so correctly and consistently. It is simply the case that the match between its information and the real world will always, to some degree, be inexact – and no reliable mechanism yet exists for restraining even egregious mismatches. Hence the fine imposed in June 2023 by a Manhattan district judge on two lawyers and a law firm after a court filing was found to contain six entirely fabricated citations generated by AI. As the judge put it, they:

abandoned their responsibilities when they submitted nonexistent judicial opinions with fake quotes and citations created by the artificial intelligence tool ChatGPT, then continued to stand by the fake opinions after judicial orders called their existence into question.[22]

What determines the accuracy of an LLM's output? In the case of my first query, ChatGPT's response closely matched a well-established pattern of evidence. In the case of my second query, the relationship between the system's inferences and actuality broke down. Yet the mechanics behind each of these results are opaque. If human consciousness consists of a constant dialogue between the mind's generative potentials and the body's sensory feedback, LLMs are more like aliens whose entire universe consists of a trillion ingested webpages. They are not only black boxes, but also inherently incurious about the truth-status of their answers. Probabilistic plausibility is all that matters.

None of this should be taken as a denigration of the achievement represented by LLMs. Indeed, the fact that such eloquent, appropriate responses can be produced by software is one of the great triumphs of modern computer science, not to mention an uneasy case study in how far it's possible to progress towards something resembling human intelligence through utterly inhuman means. But it's also a recipe for creating what the philosopher Daniel Dennett called in a May 2023 article 'the most dangerous artifacts in human history':[23] counterfeit people.

In June 2022, a Google employee called Blake Lemoine went public with the claim that an AI system he had been interacting with had become conscious.[24] It's a claim that most computer scientists deemed misconceived; yet it has been repeatedly echoed during others' interactions with the latest generation of chatbots. Why? As Dennett points out, the anthropomorphism of the 'intentional stance' makes us innately inclined to project desires and beliefs onto systems that possess neither. Thanks to the rise of systems capable of fluently and creatively producing text, images, video and audio in response to human prompts, we are rapidly entering a world in which it will be impossible to distinguish authentic from synthetic digital representations of people. These will remain *digital representations*: flesh-and-blood reality is less tractable. Online, however, both human-generated content and actual humans may soon be crowded out by echoes and imitations – matched by a new urgency around issues of truth, trust and transparency.

Artificial Intelligence is, to say the least, a vast and varied field, and startling advances are already promised across disciplines ranging from pharmacology to astrophysics; from number theory to route-finding; from interdisciplinary research to the microscopic foundations of material science.[25] As the relationship between humans and machines becomes ever closer, however, their extension and augmentation of our minds will only become more seamless and consequential. And this means that the accompanying risks will also intensify – not least that our collective grip on reality loosens at a time when facing certain facts couldn't be more vital.

Seeking alternatives to anthropomorphism

Given the ways in which our minds are already extended by technologies like writing, the fact that language itself can speak back to us when harnessed to sufficient computational power isn't, perhaps, as surprising as it might seem. Yet the fact that the machines speaking to us with such eloquence know precisely nothing about 'our' world bears repeating. Their universe is the data they've been trained upon; their language an ocean of symbols that at no point touches land. Just as AlphaZero has no idea that it is playing chess,

ChatGPT understands nothing about computers, games or academic articles. As the American philosopher and linguist Emily M. Bender argued in April 2022:

> Talking about 'training' machine learning systems is the standard terminology. I think it is always worth pausing and considering on what grounds we are talking about 'training', 'learning' and 'intelligence' in these systems; what metaphors are at play; and to what extent we are asked, as readers, to nod along to the metaphor without any clear guidance as to which aspects apply and which are merely suggestive. In the case of (large) language models . . . the 'training' involves taking a mathematical model with random mathematical parameters ('weights') and iteratively adjusting those weights in response to differences between model output and some point of comparison showing expected output . . . the primary 'training' is just next word prediction over enormous amounts of text.[26]

What the system 'knows' are domains of nested patterns and correspondences: informational graphs woven in ways that defy simplification or explanation. Countless extrapolations can be drawn from these. But their ultimate measures of success and failure – indeed, everything that connects algorithmic tokens to meanings and purposes – will remain wholly human.[27]

Similarly, perhaps the greatest danger we face isn't machines themselves so much as the purposes to which they are devoted – and the misreading of their limitations and potentials. Consider the claim that AIs pose an existential threat to humanity, typified by the following statement, issued in May 2023 by a selection of prominent industry leaders and experts:

> Mitigating the risk of extinction from AI should be a global priority alongside other societal-scale risks such as pandemics and nuclear war.[28]

It's a warning that encompasses two divergent domains: one focused on the destructive potentials of current and near-future systems; the other focused on vastly more powerful hypothetical systems. It isn't hard to populate the first of these categories with urgent and alarmingly plausible examples. From autonomous weapons to mass campaigns of disinformation, destabilization and manipulation; from the entrenching and empowering of totalitarian regimes to the destruction of vital infrastructure; from the engineering of lethal biological agents to the hacking of nuclear weapons systems; the twenty-first century contains both appalling technological possibilities and the likelihood that AIs will accelerate developments in such fields while empowering malicious actors.

Yet mitigating against such risks is above all a *human* business – and this makes the assessment of hypothetically superhuman systems troublesome. As the statement above notes, AI risk naturally sits alongside efforts to regulate potent existing technologies like nuclear weapons, biological agents and gene-editing. And what matters most in these contexts is transnational strategies for collaboration and negotiation, backed by mutual trust and scrutiny – not to mention a clear-eyed analysis of the world as it is, complete with rivalries, alliances and dysfunctions.

By contrast, speculation about the properties of machine superintelligences is a form of storytelling that, at times, reads suspiciously similarly to marketing copy. As the author Brian Merchant argued in a March 2023 column for the *LA Times*, the last few years have seen:

[the] unleashing of a science fiction-infused marketing frenzy unlike anything in recent memory. Now, the benefits of this apocalyptic AI marketing are twofold. First, it encourages users to try the 'scary' service in question – what better way to generate a buzz than to insist, with a certain presumed credibility, that your new technology is so potent it might unravel the world as we know it? The second is more mundane . . . a big driver in motivating companies to buy into automation technology is and always has been fear . . . If companies believe a labor-saving technology is so powerful or efficient that their competitors are sure to adopt it, they don't want to miss out – regardless of the ultimate utility.[29]

How *should* we set about resisting anthropomorphism while grounding discussions of existential risk, regulation and innovation in the real world? Most obviously, we need to acknowledge that an AI's outputs aren't the products of 'its' particular perspective or world-view – and that it no more has autonomous desires, intentions or ethical aspirations than any other machine.

If you ask a Large Language Model about a non-existent scientific discovery, a fictional famous figure or a bogus historical event, it will instantly be able to generate a convincing impression of a Wikipedia entry, a journalistic profile, a newspaper article – or whatever other patterns its inputs suggest are appropriate. But it doesn't and cannot think or believe anything about whatever real-world entities its inputs or outputs refer to; and nor does it have any consistent opinion, beliefs or preferences. For all its ferocious complexity, an AI is no more complex than a kettle or microwave in this respect. If you want a machine to help you write computer code, a novel or a viral genome, it will do so. If you want to automate social media trolling, the dissemination of disinformation or the mass-production of bogus academic papers, all of this can effortlessly be automated. If you tell an AI to output a score for criminality, affability, intelligence, employability – or any other arbitrary label you feel like fixing onto a weighted score – this is precisely what it will do, rapidly and with perfect confidence, no matter how pseudo-scientific your criteria or inadequate your data.[30]

Similarly, while algorithms can arrive at fairer assessments of some situations than humans by removing sources of noise and bias,[31] it is only human participation in (and scrutiny of) such systems that can anchor them to principles such as accountability or justice. This will remain true no matter how competent the systems in questions become, for algorithms can neither stand in solidarity with people nor be held answerable for their own verdicts. As the professor of legal philosophy John Tasioulas put it in a 2023 interview:

> Outcomes are not all that matter in life; processes inherently matter too. This is one reason why people support trial by jury even if, for example, trial by experts would yield better decisions. The same applies to democratic decision-making procedures. They find great value in the process of being tried by one's peers or democratic self-government, even if a hypothetical legal expert or a benevolent dictator would make better decisions. So even if AI systems could yield better decisional outcomes (a massive 'if' at this point), there is still the issue of whether, for example, they could provide adequate (justifying) explanations for them, whether they could be answerable for them, and whether their decisions would manifest the kind of reciprocity and solidarity that can be achieved when one human stands in judgment over another human.[32]

There are fascinating and immensely significant questions to be explored around the future potentials of AI and computation. But too abstract a focus on the world-changing potentials of future technologies risks becoming a distraction from the responsibilities and risks surrounding *existing* ones – not to mention the incentives and aspirations of those controlling them. As Emily Bender, Timnit Gebru, Angeline McMillan-Major and 'Shmargaret Shmitchell' put it in their 2021 paper, 'On the Dangers of Stochastic Parrots', an early and influential analysis of 'ever larger' Language Models whose conclusions have far wider significance:

> We have identified a wide variety of costs and risks associated with the rush for ever larger LMs, including: environmental costs (borne typically by those not benefiting from the resulting technology); financial costs, which in turn erect barriers to entry, limiting who can contribute to this research area and which languages can benefit from the most advanced techniques; opportunity cost, as researchers pour effort away from directions requiring less resources; and the risk of substantial harms, including stereotyping, denigration, increases in extremist ideology, and wrongful arrest, should humans encounter seemingly coherent LM output and take it for the words of some person or organization who has accountability for what is said.[33]

AIs' achievements are remarkable, and certain to become more remarkable over time. But the amazement these achievements prompt mustn't lead us to be blinded by hype to machines', and their creators', limitations – or to the fact that even the most advanced systems are still artefacts engineered by particular people for particular purposes, bearing the stamp of their creators' ambitions and limitations. As Bender concluded her 2022 essay:

> Software systems are . . . not sentient entities, and as such 'teaching' is a misplaced and misleading metaphor (as is 'machine

learning' or 'artificial intelligence') . . . When a computer seems to 'speak our language', we're actually the ones doing all of the work.[34]

What do we want from AI?

Where do we go from here? Consider an AI-driven tool like Google Translate: something so manifestly, magnificently useful that it's hard to wish it uninvented. From the point of view of reading, research and communication, what it offers would have been almost beyond belief two decades ago. From the perspective of translation as a human skill, it's at the very least disconcerting to see millions of hours of skilled labour (in the form of the billions of lines of human translations used as training sets) digested and repurposed into a proprietary service by one of the world's most powerful corporations.

On the one hand, modern machine translation emphasizes the degree to which language can be treated as data. On the other hand, it emphasizes the degree to which human words and actions are the vital ingredient without which all such services couldn't exist. As the saying goes, if you're using a free service then you are the product: a corollary to which is that if you're obliged to use enough services that treat you like a product, you may end up shrinking your self-conception accordingly.[35]

Let's return, for a moment, to the smart assistants dotted throughout my own and millions of others' homes. What are they? How and why do they do what they do? Despite appearances, they're nothing like either pets or personal helpers. They're more like the electronic sensory organs of some vast remote entity; one ceaselessly integrating our actions into patterns too intricate for any human mind to trace.

All this is both a mystery worth marvelling at and the key to our creations' inhuman power. As the philosopher and author David

Weinberger noted in a November 2021 essay for *Aeon* magazine, the field of machine learning is currently yielding ever-greater insights by abandoning preconceptions of human-like understanding in its design. But the lesson to be taken from this is not that machines can gift us certainty or negate the need for human judgement. Rather, it's that they may be able to help us acknowledge the depths of reality's complexity and *uncertainty* – so long as we don't mistake statistical inferences for truths:

> The success of our technology is teaching us that the world is the real black box. From our watches to our cars, from our cameras to our thermostats, machine learning has already embedded itself in much of our everyday lives . . . Now that we have mechanisms that stun us with a power wrung from swirls of particulars connected in incomprehensible, delicate webs, perhaps we will no longer write off those chaotic swirls as mere appearances to be penetrated.[36]

From an author's perspective there's something inherently uncomfortable about seeing a fact that has always been implicit made constantly explicit: that words and ideas are never wholly original; that we are all part of a greater ebb and flow, chaotic yet patterned to a disconcerting degree. Could an AI help me write my next book? Almost certainly, with a little coaxing. Could someone else decide they want to read a book written in my style, then get a future AI to generate it? So long as enough of my work has been analysed, quite possibly. Indeed, the person in question may prefer the opportunity to have 'me' write about any topic they like, at any length. Just so long as they accept that such a simulacrum will be both endlessly loquacious and unable to answer the most important questions of all: given everything I might have said, what would I have written about? What would I have intended, hoped and argued for – and would I have been right to do so?

The value and promise I bring to a piece of work, in other words, is that *I* chose it: that it speaks to and from the particular circumstances of a life; and that it does so via a common store of understandings, inferences and desires. Where these things matter less, automation may indeed become indistinguishable from (or preferable to) human effort. Where they matter most, it's vital to define and defend what human effort means: not just data processed or information conveyed, but thoughts and values uniquely articulated in the service of ideas, beliefs, and particular experiences of the world.

Anthropomorphism cuts both ways. To believe a machine is like us is to believe we are like machines – and, in the process, to focus upon inputs and outputs at the expense of all that lies between. Yet, like all human beings, it is only by *trying* to find words for my thoughts that I am able to work out what I mean in the first place: to fumble my way towards expressions of what it's like to be me. Countless conversations have taught me to rethink, rephrase, doubt and reread; to trust in language as a mutual, open system. As the author Ted Chiang put it in a February 2023 article for *The New Yorker*:

> If you're a writer, you will write a lot of unoriginal work before you write something original. And the time and effort expended on that unoriginal work isn't wasted; on the contrary, I would suggest that it is precisely what enables you to eventually create something original. The hours spent choosing the right word and rearranging sentences to better follow one another are what teach you how meaning is conveyed by prose. Having students write essays isn't merely a way to test their grasp of the material; it gives them experience in articulating their thoughts. If students never have to write essays that we have all read before, they will never gain the skills needed to write something that we have never read.[37]

Ultimately, this is what language *is*: a human-made and human-inhabited structure whose coherence is inextricable from thought and experience. Cut us out of the loop and, soon enough, the outputs of machines trained upon other machines' outputs degenerate towards dross.[38]

At their best, modern AIs can analyse, predict, recognize and respond to inputs with inhuman insight. They can echo any voice; generate computer code of near-flawless coherence; improvise music in any genre; conjure up art in any style; solve complex problems in geometry and biochemistry. Collaborations between skilled human users and expert systems have boundless potential: after all, Deep Blue's victory against Kasparov didn't so much herald the end of human interest in chess as mark the start of a new era of grandmasters honing their skills against supremely accomplished partners. Perhaps the greatest obstacle to achieving similarly clear-eyed collaborations elsewhere, however, is the evasion bred by the *anthropomorphic delusion*: the degree to which we are willing to attribute intentions and understanding to artificial agents which possess neither, while at the same time overlooking the all-too-human limitations of those creating, maintaining and profiting from such systems.

Ask any computer scientist what they understand, today, by the phrase Mechanical Turk and they'll point you towards Amazon's platform of that name: a source of 'artificial artificial intelligence' where humans can outsource certain informational tasks to other humans. This is because machines cannot yet perform them – or because they want to generate datasets to teach machines to perform them better. Labelling images or providing captions. Identifying better and worse translations. Correcting obvious (to human but not yet artificial eyes) breaches of common sense. Like Kempelen's marvellously misleading mechanism, it's this kind of hidden human labour that greases the wheels of global automation; and it extends far beyond Amazon's empire.

Even the most advanced AIs remain reliant to a remarkable degree upon both user feedback and the largely unacknowledged labour of thousands of human moderators. It's people who fill in the blanks, insert the understanding and test tech giants' products. It's people who censor the most offensive and disturbing content so that the rest of us don't have to endure it; who are paid a pittance to review and label datasets derived from the internet's darkest depths. As *Time* magazine headlined its January 2023 investigation into some of the workers behind ChatGPT's safety systems, 'OpenAI Used Kenyan Workers on Less Than $2 Per Hour to Make ChatGPT Less Toxic'.[39] And it's the effacement of this labour – the pretence that data is found rather than made – that most often leaves us open to deception, manipulation and delusory hope.

What do we want *from* AI? How should the lives and works its inferences are based upon – and upon which it exerts such influence – be acknowledged, protected and granted agency? These are among the most urgent questions of our algorithmic age. And answering them means thinking outside the box marked 'artificial intelligence': about the people and contexts it draws upon; its profound difference from anything approaching a human mind; and what it means to ask of automated systems questions worth answering.

CHAPTER 10

Superintelligence and doubt:
The delusion of machine perfection

Just before 10pm on 18 March 2018, a forty-nine-year-old woman called Elaine Herzberg was pushing her bicycle across a four-lane road in the city of Tempe, Arizona, when she was struck by a Volvo SUV. Half an hour later she was pronounced dead at a local hospital.

Herzberg was not on a designated pedestrian crosswalk at the time of the accident. It was a dark night, and her judgement may have been impaired by drugs. According to the National Transportation Safety Board's report into the accident, however, the woman in the driver's seat, Rafaela Vasquez, might reasonably have been expected to see Herzberg and act in time to avoid her.[1] Vasquez hadn't done this because, her mobile phone records suggested, she was streaming an episode of 'The Voice' rather than paying full attention to the road (Vasquez has subsequently argued that she was listening to, but not watching, the programme).[2]

Every fatal accident is of terrible significance. But this particular accident was also significant in another sense. If Rafaela Vasquez had noticed Elaine Herzberg in time, her life might have been saved. But Vasquez wasn't the one controlling the car. Instead, it was being operated by an autonomous driving system created by the ride-share tech company Uber. Vasquez was supposed to watch the road at all times and intervene if needed. Instead, Herzberg suffered the tragic distinction of becoming the first pedestrian to be killed by a vehicle controlled by a computer rather than a human.

When technological dreams meet the real world

Few topics are more enticing to armchair ethicists than autonomous vehicles, not least because they're a rare everyday example of computers wielding potentially lethal force near humans. So-called 'trolley problems' (named after a 1967 paper in which the philosopher Philippa Foot used the thought experiment of a runaway tram to explore moral intuitions around harm and duty)[3] are staples of heated debates about, for example, whether a vehicle should be programmed to avoid a group of people even if it means driving into a lone pedestrian; or whether it should privilege its occupants' safety over that of other road-users.

There's plenty of philosophical interest in thinking up confounding scenarios. Saving three lives at the cost of one may be the right thing to do, but would *you* buy a car designed to drive itself off a cliff if the only alternative was hitting several jaywalkers? In practice, however, the development and regulation of autonomous vehicles is a less theoretical business; one whose 'accidental revelations' are more likely to involve unforeseen circumstances and sheer bad luck than failures of philosophical logic. Here, for example, is the summary of the 'probable cause' of the accident of 18 March 2018 ultimately arrived at by the NTSB.

> The probable cause of the crash in Tempe, Arizona, was the failure of the vehicle operator to monitor the driving environment and the operation of the automated driving system because she was visually distracted throughout the trip by her personal cell phone. Contributing to the crash were the Uber Advanced Technologies

Group's (1) inadequate safety risk assessment procedures, (2) ineffective oversight of vehicle operators, and (3) lack of adequate mechanisms for addressing operators' automation complacency – all a consequence of its inadequate safety culture. Further factors contributing to the crash were (1) the impaired pedestrian's crossing of N. Mill Avenue outside a crosswalk, and (2) the Arizona Department of Transportation's insufficient oversight of automated vehicle testing.[4]

As the report's conclusions emphasize, the probable cause was that the vehicle's operator, Rafaela Vasquez, failed to pay sufficient attention. But why was she inattentive? Partly because of distraction and poor judgement. But partly, also, because 'automation complacency' – a phrase describing the human tendency gradually to become inattentive when monitoring an automated system – was neither sufficiently considered in this particular system's design, nor mitigated against in protocols for overseeing its operators. To borrow a phrase from the AI researcher Madeleine Clare Elish, Vasquez in effect acted as a 'moral crumple zone', bearing the brunt of the ethical and legal responsibility for the overall system's dysfunctions.

> While the crumple zone in a car is meant to protect the human driver, the moral crumple zone protects the integrity of the technological system, at the expense of the nearest human operator.[5]

Why didn't the car manage to stop itself? If it had been a non-automated version of the vehicle, it might have done. Like most modern cars, Volvo SUVs incorporate an automatic braking system designed to prevent collisions. But this system had been switched off so that Uber's software, radar, lidar, cameras and GPS could fully control the vehicle. These systems did indeed identify *something* in the road ahead. But the AI controlling the car didn't classify it as a jaywalking pedestrian pushing a bicycle, because such a category didn't exist in its code. Instead, it effectively treated Elaine Herzberg as a series of different possible objects flashing into and out of existence. Here is the report's account of the calculations made by the AI prior to impact.

> The ADS [Automated Driving System] first detected the pedestrian 5.6 seconds before impact . . . The system never classified her as a pedestrian – or correctly predicted her path – because she was crossing N. Mill Avenue at a location without a crosswalk, and the system design did not include consideration for jaywalking pedestrians. The ADS changed the pedestrian's classification several times, alternating between vehicle, bicycle, and other. Because the system never classified the pedestrian as such, and the system's design excluded tracking history for nonpersisting objects – those with changed classifications – it was unable to correctly predict the pedestrian's path.[6]

The AI, in other words, was both incredibly capable within certain parameters and perilously incapable outside them. Unanticipated circumstances exposed these limitations – which is, of course, the whole point of testing. Unlike most software tests, however, the consequence was a literal rather than metaphorical crash.

A series of recommendations attempting to redress the failures behind the crash duly followed: that the National Highway Traffic Safety Administration should require those intending to test automated driving systems to submit self-assessment reports, and establish a process for the ongoing evaluation of these reports; that the state of Arizona should require applications for testing ADS-equipped vehicles, including details of risk management and countermeasures, as well as establishing an expert task group to evaluate these; that the American Association of Motor Vehicle Administrators should inform other states about the circumstances of the Tempe, Arizona, crash and encourage them to take similar steps to Arizona; and that Uber should complete the implementation of a safety management system for ADS testing which 'at a minimum, includes safety policy, safety risk management, safety assurance, and safety promotion.'[7]

It's worth listing these (notably unexciting) recommendations because none of them have anything to do with what was going on 'inside' the computer system being tested. Instead, they concern the culture and processes surrounding its development and deployment. They don't specify the details of a perfectly safe system, because such a system doesn't exist. They don't fixate on the particular circumstances of the accident, because it's in the very nature of accidents that the next one will be caused by different circumstances. Robust safety and assurance protocols necessarily begin with the assumption that unpredictable and undesirable events *will* occur – and thus that the last thing anyone involved can afford is to hope that tech will magically banish them. They are unexciting by design: cautious, unsensational; a low-key form of automation aimed at restraining rashness and overpromising.

They're also bound up with human oversight, debate and accountability. As the professor of science and technology studies Jack Stilgoe puts it in his 2019 book *Who's Driving Innovation?*, which opens with an in-depth analysis of the crash:

> When technological dreams meet the real world, the results are often disappointing and occasionally messy. Policymakers are seduced by the promise of new technologies, which arrive without instructions for how they should be governed. It is all too common for regulation to be an afterthought. In the world of aviation, it's called a tombstone mentality . . . lessons are learned and rules

rewritten in grim hindsight. In Arizona, policymakers allowed a private experiment to take place in public, with citizens as unwitting participants. It ended badly for everyone involved.[8]

Greater scepticism should have been shown before this particular 'private experiment' was allowed to play out on public roads. But this doesn't mean that self-driving cars are inherently unsafe, that they can't be tested in real-world situations, or indeed that moving towards a world in which vehicles are primarily controlled by computers is a bad thing. Rather, as Stilgoe emphasizes, it shows that 'technological dreams' are an unsound basis for public policy – and that contrasting infallible machines to inept humans has little to do with real life. As he goes on to suggest:

> Humans debase themselves when presented with a story of what [the author] David Nye calls the 'technological sublime'. Within this story, the general public are cast in a very limited role: catch up, adapt, accept the inevitable . . . When humans compare themselves to an idealised machine, the imperfections stand out. Even though humans are the only control devices ever to have successfully driven cars, some are ready to make the argument that driving should be banned.[9]

What is an 'idealized machine'? It's one, among other things, that's able to operate in ideal conditions. In the case of most modern AIs, this means the precisely predictable realm of a game like chess, or a physical environment engineered to reproduce this regularity. Such systems may indeed be preferable to humans in these contexts: a factory assembly line; a warehouse designed around machines' affordances; a customer support service aimed rapidly to address a set range of queries. But it's worth noting that machine-optimized environments have little in common with conditions that human workers find engaging, meaningful and dignified; and that there's a vast difference between automating dangerous, dull or demeaning labour and eroding the skill, satisfaction and autonomy of those jobs that remain.

Another way of putting this is that the principles of human-centred design require a certain inefficiency and openness, not to mention an interest in supporting rather than supplanting human agency.[10] Compare the clarity of a goal like winning a game of chess, or driving a car around a closed circuit, with the 'goal' of safely and pleasurably navigating a busy city. As Stilgoe points out, humans are still the only 'control devices' able reliably to master this last objective, largely because it's actually an endless series of overlapping tasks.

Similarly, the most reliable and cost-effective ways of making driving safer still involve low-tech innovations like seatbelts, automatic braking and improvements to road design. That's because these 'dumb' features are both robust in the face of unpredictable circumstances and beneficial to all kinds of humans – rich, poor, jaywalking, cycling – whether or not these humans are doing what a manufacturer wishes or an algorithm has foreseen. It may not be particularly profitable to enforce standards, craft hospitable civic environments or provide a diversity of transport options. Yet it's measures like these that do most to make a society liveable.

Expecting too much from machines

What is the current state of the art when it comes to autonomous vehicles? AIs are already able to assist with many aspects of driving: matching speeds and staying in lane on highways; braking, parking and monitoring other vehicles' movements. The Society of Automotive Engineers has developed a scale to measure different levels of autonomous driving, with zero indicating none and five indicating full real-world autonomy. As of early 2023, Tesla's 'autopilot' system qualified for SAE level two – 'partial driving automation' – while Mercedes-Benz was the only automaker approved to produce commercial systems at SAE level three, with its Drive Pilot system capable of 'conditionally automated' driving on certain tasks.[11] By contrast, the realm of autonomous taxis is more developed, with firms such as Waymo (formerly the Google self-driving car project) offering a fully driverless service in select cities such as San Francisco.[12]

Even if or when widespread autonomy arrives, however, the navigation of many real-world environments will continue to require human participation. As an Italian professor of computer science once put it to me during a conference near Rome, how could any vehicle hope to navigate that city's busy, narrow streets without its driver making regular (and vigorous) disambiguating gestures? The world cannot simply be wiped clean of its history. Or rather, the price of erasing this history can't be assessed in isolation from other concerns.

Consider the ways in which a gamut of 'smart' technologies are already bringing with them subtler and more significant changes than any enhancement to convenience. The following is a story from February 2023, as told by the author and activist Cory Doctorow.

> The masked car-thieves who stole a Volkswagen SUV in Lake County, IL didn't know that there was a two-year-old child in the back seat – but that's no excuse. A violent car-theft has the potential to hurt or kill people, after all. Likewise, the VW execs who decided to nonconsensually track the location of every driver and sell that data to shady brokers – but to deny car owners access to that data unless they paid for a 'find my car' subscription – didn't foresee that their cheap, bumbling subcontractors would refuse the

local sheriff's pleas to locate the car with the kidnapped toddler. And yet, here we are. Like most (all?) major car makers, Volkswagen has filled its vehicles with surveillance gear, and has a hot side-hustle as a funnel for the data-brokerage industry . . . The local sheriff called Volkswagen and begged them to track the car. VW refused, citing the fact that the mother had not paid for the $150 find-my-car subscription after the free trial period expired. Eventually, VW relented . . . but not until after the stolen car had been found and the child had been retrieved.[13]

It's a story with the character of Virilio's 'accidental revelations', laying bare dynamics that go unremarked in everyday circumstances only to surface at a moment of crisis. And it emphasizes how even non-autonomous modern vehicles, though they may seem similar to every other car from the last few decades, have begun to conceal all-too-new affordances. What will it mean to populate the places we live, love and work with countless machines that serve as mobile monitoring devices, bristling with proprietary software and opaque licensing agreements? What *should* it mean to hold accountable those profiting from this? As the author Nicholas Carr put it in his 2014 book *The Glass Cage*:

> When an inscrutable technology becomes an invisible technology, we would be wise to be concerned. At that point, the technology's assumptions and intentions have infiltrated our own desires and actions. We no longer know whether the software is aiding us or controlling us. We're behind the wheel, but we can't be sure who's driving.[14]

The problem is not that all forms of automation diminish human agency, or that its promises are inherently hollow. Various forms of driver assistance will undoubtedly make roads far safer over the coming decades, just as the places in which we live and work will undoubtedly evolve alongside technologies' affordances. Rather, the issue is how *delusions of machine perfection* efface a vital, ongoing negotiation that every twenty-first-century society ought to facilitate: between its citizens and their technosocial environment; between humanity's needs and technology's desires.

To return to the Douglas Adams essay this book began with, invisible technologies aren't simply those that work as intended. They're also a kind of magic trick: an engineered bait-and-switch that presents certain outcomes as a *fait accompli*. Want to know where your car is at all times? Want to reduce the cost of your insurance? You'll need take out a subscription and submit to surveillance. Want to stay employed? You'll need to swap personal autonomy for algorithmic optimization. In the words of Sherry Turkle, we are being encouraged to 'expect more from technology and less from each other'.[15] And it's a mismatch that risks squandering both our own and our creations' divergent potentials.

Rethinking Artificial Intelligence

As the previous chapter explored, it's hard to overstate either the significance of AI or the degree to which it's already interwoven with our lives. From search engines to social media and thermostats, from shopping recommendations to satellite navigation and fraud detection, some form of machine learning increasingly sits between us and every single automated system we encounter. Ours is an informational age whose raw materials are useful and comprehensible *only* through technologies of ever greater complexity – and AI is the driving force behind this. Yet there remain several troubling things about both the term 'Artificial Intelligence' and the invidious comparisons it invites between minds and machines.

For a start, the analogy embedded within 'AI' enacts an enduring hubris bound up with the field's origins. In 1955, a group of American scientists announced an event they believed might transform the world. The title they chose was The Dartmouth Summer Research Project on Artificial Intelligence. This was the first time the term AI had been used, and the expectations attending it were explicitly immodest. As the project's proposal put it:

> The study is to proceed on the basis of the conjecture that every aspect of learning or any other feature of intelligence can in principle be so precisely described that a machine can be made to simulate it. An attempt will be made to find how to make machines use language, form abstractions and concepts, solve kinds of problems now reserved for humans, and improve themselves. We think that a significant advance can be made in one or more of these problems if a carefully selected group of scientists work on it together for a summer.[16]

In the more than six decades since then, remarkable progress has indeed been made in some of these areas: language usage, problem-solving, self-improvement. Yet the creation of machines capable of abstraction and concept formation remains in its infancy – while the possibility of describing and replicating 'every aspect of learning or any other feature of intelligence' still seems distant. The supposition that human mental life might soon be described and replicated has proved naive in every sense: a demonstration not so much of machines' potentials as of the human tendency to underestimate our own complexity.

In effect, AI's foundational document envisioned intelligence as a fungible attribute that could be exported from the human to the machine realm via a sufficiently meticulous descriptive process.

It's a vision akin to Cartesian dualism, defining 'intelligence' as an abstraction distinct from emotion, embodiment, socialization and subjectivity. Even taken on its own terms, AI's history embodies anything but a straightforward progression towards such an ideal. To list just a few of its intricacies, modern AI encompasses computational approaches to planning, reasoning, search, knowledge representation and machine learning; while this latter subfield encompasses probabilistic learning, decision trees, genetic algorithms and deep learning. There have been several periods of stagnation, overpromising and under-delivery since the 1950s, followed by revelations and fresh promises. The last decade in particular has seen stunning progress. But there remains ongoing debate around different approaches' strengths and limitations, how best to combine them, and what new techniques and insights are yet be found.

As the machine intelligence experts Gary Marcus and Ernest Davis explore in their 2019 book *Rebooting AI*, from which the above taxonomy is drawn, it is above all *deep learning* that lies behind recent breakthroughs. But to depict its data-hungry brilliance as a path towards the holy grail of human-like Artificial General Intelligence (AGI) is, in their view, misguided and short-sighted. This is because deep learning is *greedy*, *opaque* and *brittle*. It requires immense amounts of data to arrive at valid inferences; it achieves its results in inscrutable ways; and its models can be 'broken' by exposure to novel situations.

One of the pioneering firm DeepMind's first breakthroughs, for example, was the unveiling in 2013 of a deep learning model able to play a number of classic video games. This approach was groundbreaking in its use of deep reinforcement learning to improve performance based upon little information beyond the pixels and scores generated by each game. Within the constraints of its starting conditions, the model achieved superhuman competence. This competence, however, tended to vanish in the face of even tiny alterations to these starting conditions.

> It is important to recognize that, in the term deep learning, the word 'deep' refers to the number of layers in a neural network and nothing more. 'Deep,' in that context, doesn't mean that the system has learned anything particularly conceptually rich about the data that it has seen. The deep reinforcement learning algorithm that drives DeepMind's Atari game system, for example, can play millions of games of Breakout, and still never really learn what a paddle is . . . In Breakout, the player moves the paddle back and forth along a horizontal line. If you change the game so that the paddle is now a few pixels closer to the bricks (which wouldn't bother a human at all), DeepMind's entire system falls apart.[17]

What follows from this? AIs have improved vastly since 2013, and systems like LLMs can now fluently handle multiple frames of reference. Marcus in particular has continued to make the case, however, that major risks associated with fragility remain – not least the 'hallucinatory' tendency to fabricate plausible outputs in the absence of reliable data – and that engaging hopefully with the future of AI means incorporating insights from the multiplicity of tools, structures and forms of 'world knowledge' that characterize biological consciousness.[18]

As the first half of this book explored, human minds conjure actionable apprehensions of the world through a combination of top-down and bottom-up mechanisms: through rich narratives of causation and interrelation; and through an evolved sensibility driven by empathy, imagination and curiosity. Much of this is implicit in the phrase 'common sense' which, Marcus and Davis suggest, invokes nothing less than an innate set of theories about the properties of the world. This in turn suggests a future direction for AI that's not so much about superhuman statistical extrapolations as the incremental, imperfect encoding of different kinds of comprehension.

> Start by developing systems that can represent the core frameworks of human knowledge ... Develop powerful reasoning techniques that can deal with knowledge that is complex, uncertain, and incomplete and that can freely work both top-down and bottom-up. Connect these to perception, manipulation, and language. Use these to build rich cognitive models of the world. Then finally the keystone: construct a kind of human-inspired learning system that uses all the knowledge and cognitive abilities that the AI has; that incorporates what it learns into its prior knowledge; and that, like a child, voraciously learns from every possible source ... [19]

Far from embodying an accelerating ascent to superhuman heights, this version of automation's future looks more like a gradual deepening of humanity's self-knowledge: a reverse engineering of our biological inheritance in all its multiplicity.

Does such an account bear any relation to what will actually happen? As ever, the truth is likely to prove more complex and compromised. In ethical and imaginative terms, however, it's a vision whose explicit invocations of childhood and learning suggest a humane way of contemplating our future relationships with intelligent (not to mention super-intelligent) machines: as parents guiding them, responsive and responsible to principles of mutual care. Equally importantly, it acknowledges just how far we have to go when it comes to coding anything resembling our own modes of learning or understanding – not to mention how careful we need to be not to treat mindlessly brilliant optimization as an adequate aspiration for either our own or our creations' future.

Encoding doubt into machines

There are deep contradictions in many of the stories twenty-first-century societies tell about technology: that it's neutral, yet can confer meanings and purposes upon us; that it can be understood by analogy to our own capabilities, yet its brilliance is inhuman and incomprehensible; that human beings are irredeemably flawed, yet a technocratic class can solve our collective crises. The common thread here is an impatience with ambiguity and uncertainty – and an unwillingness to accept that, collectively, we both bear responsibility for our creations and are the sole source of the values they enact.

All of this is especially urgent when it comes to questions of automation, because it is here that we must define the goals and priorities our creations will pursue at an ever-accelerating rate on our behalf. As Stuart Russell puts it at the start of *Human Compatible*:

> The problem is right there in the basic definition of AI. We say that machines are intelligent to the extent that their actions can be expected to achieve their objectives, but we have no reliable way to make sure that their objectives are the same as our objectives.[20]

In principle, he suggests, machines can be defined as beneficial to humans 'to the extent that their actions can be expected to achieve our objectives'. In practice, 'our' objectives are inherently plural, confused and conflicting. Over eight billion humans live on this planet, not to mention countless creatures capable of pleasure and suffering, with trillions more of each potentially populating the epochs ahead. What might it mean to make and maintain machines that are 'beneficial' in such a context? Above all, Russell suggests, the crucial element missing from most discussions of technology's future is one that explicitly acknowledges this multiplicity: *doubt*.

> The difficult part, of course, is that our objectives are in us . . . and not in the machines. It is, nonetheless, possible to build machines that are beneficial in exactly this sense. Inevitably, these machines will be uncertain about our objectives – after all, we are uncertain about them ourselves – but it turns out that this is a feature, not a bug (that is, a good thing and not a bad thing). Uncertainty about objectives implies that machines will necessarily defer to humans: they will ask permission, they will accept correction, and they will allow themselves to be switched off.[21]

As Russell points out, almost every real-world decision is computationally complex to such a degree that it cannot be perfectly solved. In a 'solved' board game like checkers, every single move in every possible game has been calculated – and this means that it is possible to say at any stage of any game what an ideal move would be. If two computers play checkers perfectly against one another, the result will always be a draw. By contrast, chess is sufficiently complex that – although AIs are remarkably adept at calculating moves likely to lead to victory – perfection is unlikely to arrive this side of several processing revolutions.[22]

Even chess, however, is absurdly simple compared to everyday life. What would it mean to come up with a perfect answer to a question like 'What should I have for lunch?' Resolving this would, at a minimum, entail simulating every possible future resulting from every conceivable option, not to mention coming up with a reliable measure of satisfaction for each. Needless to say, this is impossible; and this will remain true no matter how much computational power you possess. No matter how remarkable machines may become, neither they nor we will ever approach perfect rationality.

None of this means that better actions cannot be distinguished from worse. Indeed, modern AI is – like us – inherently probabilistic in its approach to patterns and predictions. But it *does* mean that unwarranted certainty poses a profound problem in the context of systems operating beyond human comprehension, whether or not these systems possess understanding in any human sense. One thought experiment sometimes invoked in this context is the philosopher Nick Bostrom's hypothetical 'paperclip maximiser'. Here is one of his earliest descriptions of it, in a 2003 paper on the ethics of advanced artificial intelligence:

> It also seems perfectly possible to have a superintelligence whose sole goal is something completely arbitrary, such as to manufacture as many paperclips as possible, and who would resist with all its might any attempt to alter this goal.[23]

A machine superintelligence might, in other words, accidentally end up pursuing a carelessly defined objective ('please make as many paperclips as possible') with such inhuman brilliance and devotion that the side-effects would be catastrophic. Such an AI might decide that the existence of humans – or, indeed, of anything apart from paperclip-manufacturing facilities – was undesirable, and might thus pre-emptively eliminate all life on Earth. As Bostrom noted in a 2014 interview, this in turn points to a more general problem: that any number of different goals could in theory entail disastrous unintended consequences if pursued with superhuman diligence.

> It's not that most of these goals are evil in themselves, but that they would entail sub-goals that are incompatible with human survival . . . So a big part of the challenge ahead is to identify a final

goal that would truly be beneficial for humanity, and then to figure out a way to build the first superintelligence so that it has such an exceptional final goal.[24]

Seeking goals compatible with human thriving is all very well. But we shouldn't need hypothetical scenarios to remind us that there are dangers inherent to pursuing *any* aspiration regardless of its cost — and that seeking to pin down an 'exceptional final goal' for superhuman intelligences may, by definition, be an unwise idea. Whether you're a mind or a machine, the objectives you aspire towards and the conclusions you arrive at can only be as good as the information and assumptions you begin with. And this means that there will *always* be more, in the long term, that you need to consider.[25]

Indeed, many of the most significant ethical hazards we face when it comes to technology actively entail attempts at identifying desirable or undesirable far-future scenarios — because reasoning *backwards* from these introduces an unwarranted degree of simplicity and certainty into present debates. To echo Russell's argument, perhaps the most important thing both brilliant machines and powerful people need to bear in mind is that constructive doubt is a necessary feature of any decent strategy; and that being willing to change your mind (or allowing yourself to be switched off) may be the most important principle of all.

If, for example, you begin by assuming that the creation of beneficent superintelligent AI is both civilization's supreme task and a potential solution to all its problems, then anything that lies along this path may become an acceptable cost — while past and present injustices shrivel into insignificance. If you decide that unilateral actions by wealthy individuals are the most effective way of bringing about positive planetary change, then the greater good is served by clearing all barriers to such actions. If you believe that economic growth and resource extraction *must* be indefinitely sustained to solve the world's problems, any objections to this position can automatically be dismissed as unrealistic or regressive.[26] The more you're convinced your position is watertight, the more you may feel entitled to ignore everyone else. And the more likely you are to succumb to precisely the errors machine learning systems can nudge us towards: fragility and reality-denial; the conflation of data with truth; the denigration of human capabilities, competence and potentials.

In a sense, it all comes down to efficiency – and how ready we are for *any* values to be relentlessly pursued on all our behalfs. Writing in *Harper's* magazine in 2013, the essayist Thomas Frank considered the panorama of fast-food chain restaurants that skirts many American cities. Each outlet is a miracle of modular design, resolving the production and sale of food into an impeccably optimized operation. Yet, Frank notes, the system's success on its own terms comes at the expense of all those things left uncounted: the skills it isn't worth teaching a worker to gain; the resources it isn't cost-effective to protect.

> The modular construction, the application of assembly-line techniques to food service, the twin-basket fryers and bulk condiment dispensers, even the clever plastic lids on the coffee cups, with their fold-back sip tabs: these were all triumphs of human ingenuity. You had to admire them. And yet that intense, concentrated efficiency also demanded a fantastic wastefulness elsewhere – of fuel, of air-conditioning, of land, of landfill. Inside the box was a masterpiece of industrial engineering; outside the box were things and people that existed merely to be used up.[27]

Society has been savouring the fruits of automation since well before the industrial revolution – and the resulting gains in everything from leisure and wealth to productivity and health have been vast. As agency passes out of our hands in the name of various efficiencies, however, the losses outside these boxes don't evaporate into non-existence. Seen from inside a car, urban sprawl may seem as inevitable as vehicle ownership. Seen from the outside, however, hundreds of miles of freeway clogged by thousands of cars may look more like Frank's 'fantastic wastefulness': the externalities of outmoded affordances.

Similarly, the more we try to turn people into deterministic components within machine-readable environments, the more these people are likely to look incompetent, inadequate or superfluous. In a machine-perfected road system, no pedestrian will ever cross a road except at a crossing. Exceeding the speed limit will be physically impossible. Those who transgress against other regulations will be identified and restrained inside their autonomous vehicles. In the name of order and efficiency, law enforcement will pool the surveillance capabilities of every vehicle and traffic camera, every mobile phone and wearable device. The prize of perfect safety will be pursued hand in hand with the perfect denigration of human autonomy. The system will be optimized, and so will we – if by optimization you mean the abandonment of any open interrogation of aims, values and freedoms.[28]

What degrees of agency and accountability should be accorded to an autonomous vehicle's owners; to its manufacturers; to law enforcement; to pedestrians? What should we aspire towards when it comes to the future of transport, urban planning, work and leisure? These are vital questions for governments, citizens and regulators; for engineers and manufacturers; for drivers and passengers; for artists and architects. They will be answered in different ways in different nations, at different times. But they can only begin to be addressed rigorously if they are asked honestly and repeatedly in the light of new knowledge – and if a caricature of human fallibility isn't turned into a self-fulfilling prophecy.

Superhuman deciders may, some day, conduct these investigations alongside us. For now, however, even the smartest AI will relentlessly follow its code once set in motion. And this means that if we hope meaningfully to debate the adaptation of a human world into a machine-mediated one, this debate must take place at the design stage – or not at all. As the next chapter explores, our great task is not to create inherently ethical machines, or to outsource every responsibility to automation. Rather, it is to ask what it means for us to be at once parents and offspring to our creations: mutually dependent; divergent in our capacities; tasked with teaching them a world.

CHAPTER 11

Towards a new ethics of technology:
The delusion of divine data

In philosophy, an *epiphenomenon* is a secondary effect: something that occurs as a result of a primary phenomenon, but that isn't a cause of that phenomenon. When an old-fashioned locomotive releases some of the steam in its boiler in order to sound its whistle, the resulting noise is the product of the pressure inside the boiler, not its cause. Someone who assumed a locomotive's whistle was causing it to move would be confusing cause and effect: misinterpreting a mere symptom as something primary.[1]

Similarly, some philosophers and psychologists have suggested that consciousness itself is a kind of epiphenomenon. Like steam whistling out of an engine, it is a by-product of physical processes but not a cause of them. At best, according to this view, consciousness is a kind of storytelling about physical processes that have already played out elsewhere – and all forms of human behaviour can thus be explained in terms of physical rather than mental events. As one of the fathers of behavioural psychology, John B. Watson, put it in an influential lecture in February 1913 to the New York branch of the American Psychological Association at Columbia University:

> I believe we can write a psychology . . . [and] never use the terms consciousness, mental states, mind, content, introspectively

verifiable, imagery, and the like . . . It can be done in terms of stimulus and response, in terms of habit formation, habit integrations and the like. Furthermore, I believe that it is really worth while to make this attempt now . . . Once launched in the undertaking, we will find ourselves in a short time as far divorced from an introspective psychology as the psychology of the present time is divorced from faculty psychology.[2]

Watson's aspiration was to reshape psychology as an objective science: to understand human behaviour entirely in terms of observable facts. Scientists, he suggested, should aim at wholly material explanations of 'even the more complex forms of behaviour, such as imagination, judgment, reasoning, and conception.' This would take some time because 'our minds have been so warped' by the habit of subjectivity – but everything would eventually yield to the purity of behavioural analysis.

Although Watson was speaking over a century ago, the notion that science and technology might become a subjectivity-free method for achieving universal understanding has retained its currency. Here's the late physicist Stephen Hawking setting out his version of this view in 2011.

Why are we here? Where do we come from? Traditionally, these are questions for philosophy, but philosophy is dead. Philosophers have not kept up with modern developments in science. Particularly physics.[3]

Leaving aside Hawking's reasonable claim that many philosophers would do well to learn more about physics, the essence of his argument is that philosophical debate is a kind of epiphenomenon when it comes to science and technology: a narration generating clouds of verbal steam while accomplishing nothing. In effect, this line of argument answers many of the questions in this book with a shrug: *who cares?* While it's endlessly possible to discuss technology in terms of values, affordances and purposes, if Watson and Hawking are correct then such discussions are immaterial in the most literal of senses.

Thanks to Hawking's fame (and bluntness), his comments attracted considerable attention at the time.[4] But their fundamental flaw is easy enough to identify. The claim that science has no need of philosophy is *itself* philosophical rather than scientific. It is simultaneously a claim about the scope of science, how it should be practised, and what it means adequately to address certain fundamental questions. Yet none of these issues can be resolved from within any particular scientific paradigm, because the question of *how* and *why* a particular paradigm is, or isn't, the right way to think about them inherently lies elsewhere. As the philosophers Rani Lill Anjum and Stephen Mumford put it in their 2018 book *Causation in Science*:

> if our task is to understand what the correct norms of science are, the answer clearly cannot come from within science itself, for any such answer would be question begging . . . there is a philosophy within science whether one likes it or not. Science rests upon philosophical assumptions, including metaphysical ones. These assumptions cannot be proven by science itself, but only assumed.[5]

As Anjum and Mumford point out, the business of conducting empirical investigations inexorably relies upon a host of non-empirical assumptions: that there is such a thing as an external world that it's possible to make observations about; that observations made in particular places at particular times are relevant to other times and places; that we can legitimately treat these observations as being attributable to a particular phenomenon, rather than (say) to our imaginations or blind chance; that interventions and predictions are meaningful ways of investigating reality; and so on. In the end, they note:

> there are two types of scientists: those who are aware of science's philosophical underpinnings and those who are not. Only if we are aware of what we have assumed are we able to reflect critically upon its various aspects and ask whether it really gives us a sound foundation.[6]

If only we could gather enough data . . .

Why is all of the above worth going into in such detail? Because one of the most fundamental hurdles to be overcome in any discussion of the purposes and values embedded in technology is the claim that all we *really* need to do is gather sufficient data, then submit it to sufficiently powerful and impartial scrutiny. Chris Anderson, then editor of *Wired* magazine, put it like this in a 2008 piece titled 'The End of Theory'.

> This is a world where massive amounts of data and applied mathematics replace every other tool that might be brought to bear. Out with every theory of human behavior, from linguistics to sociology. Forget taxonomy, ontology, and psychology. Who knows why people do what they do? The point is they do it, and we can track and measure it with unprecedented fidelity. With enough data, the numbers speak for themselves.[7]

Anderson was setting out to be provocative. Even taken on their own terms, however, his claims make little sense. Consider the phrase 'the numbers speak for themselves' – something that, in this context, can neither be literally nor metaphorically true. In the literal sense, the whole point of numbers is that they do not and cannot speak for themselves: that to discuss the meaning of any measurement is to tell some kind of story *about* it. In the metaphorical sense, numbers are said to speak for themselves if the story they imply is so self-evident that it requires no explanation. But the politics of (say) vaccination, health beliefs and misinformation suggest that, at best, it's naive to think this; and that, at worst, leaving numbers to speak 'for themselves' is akin to leaving a hospital unlocked and expecting people to arrange their own healthcare.

Ultimately, there is no limit to the number of different narratives any amount of data can be made to serve; and no ultimate form of arbitration between these narratives to be found within data itself. If something seems obvious, it's because certain attitudes, expectations and assumptions make it so. And these will invariably relate to someone, somewhere, operating from within a particular frame of reference. Nothing is ever obvious to everyone, while perhaps the greatest irony of triumphalist rationalism is that it relies upon precisely the kind of belief it claims to oppose: the faith that one mode of inquiry can manifest authority across every domain. As the philosopher Mary Midgley puts it in her 2018 book *What is Philosophy For?*:

> Instead of seeing the physical sciences as real, but limited, sources of knowledge about material facts, we are now called on to revere them as the metaphysical source of all our knowledge . . . In fact this whole reductive programme – this mindless materialism, this belief in something called 'matter' as the answer to all questions – is not really science at all. It is, and has always been, just an image, a myth, a vision, an enormous act of faith. As Karl Popper said, it is 'promissory materialism,' an offer of future explanations based on boundless confidence in physical methods of enquiry.[8]

'Who knows why people do what they do?' Chris Anderson asks. The answer, of course, is *nobody* – not exhaustively. By their very nature, explanations can never be final. There are always further ways of framing a question: further assumptions to interrogate, deeper uncertainties to acknowledge. It's among the greatest strengths of modern science that it embraces and seeks incrementally to refine our grasp of such uncertainties; and it is one of the greatest legacies of behaviourism that it has disciplined us to test our beliefs about human minds against empirical evidence. But this doesn't mean that gathering more and more information can gradually take us beyond theories and subjectivity. It simply pushes our theories and assumptions back a level: towards questions of what we should be measuring, how, and why; towards a negotiation between incommensurable rights, priorities and values.

All of this brings us to this chapter's subtitle: *the delusion of divine data*. For there is a *third* sense in which the sentiment 'with enough data, the numbers speak for themselves' can be read. The first three words of Anderson's statement are significant because what they evoke is nothing less than a quasi-religious moment of revelation: a mystical point beyond which sufficient quantities of data somehow become self-sufficient and self-interpreting. When you gather 'enough' of them, numbers can apparently take on a life of their own. This is Popper's 'promissory materialism' writ large, with a flavour of Kurzweil's Singularity: a vanishing point beyond which sheer scale dissolves all questions into answers.

Like several of the other delusions discussed in this book, such a claim is seductive partly because it gestures towards important truths: towards the power, novelty and strangeness of new technological horizons. But it also abandons analysis and critical engagement at precisely the point both of these things are most precious. It's no good throwing your hands up in the air the moment exponential complexities arrive on the scene. Rather, in order to address the staggering power of technologies like big data, applied mathematics and machine learning, we must redouble our commitment to questions that *inherently* lie beyond any empirical verdict. What are our own foundational assumptions? What demands do they make of us? Upon what basis are we prepared to accept *any* claim as intellectually or ethically forceful?

Comparing different ethical systems

Numbers may not be able to tell their own stories, but Anderson is surely right to suggest that data-gathering and analysis on a superhuman scale present profound challenges to the stories through which we seek to understand the world. Indeed, our relationships with various exponentially increasing forms of complexity are central to present challenges. In her 2016 book *Technology and the Virtues*, Shannon Vallor coins the term 'technosocial opacity' to capture this point.

> Our present condition seems not only to defy confident predictions about where we are heading but even to defy the construction of a coherent narrative about where exactly we are. Has the short history of digital culture been one of overall human improvement or decline? On a developmental curve, are we approaching the next dizzying explosion of technosocial progress as some believe, or teetering on a precipice awaiting a calamitous fall . . . Our growing technosocial blindness, a condition that I will call acute technosocial opacity, makes it increasingly difficult to identify, seek, and secure the ultimate goal of ethics – a life worth choosing, a life lived well.[9]

If, in such a context, we wish to invoke such ideas as 'ethics' and 'purpose', where can we look for guidance as to what they mean?

This question concerns what's known as *meta-ethics*: the discussion of how we define fundamental concepts such as right, wrong, goodness and morality. What does it mean to offer a coherent, compelling account of ethics for our times? I believe that the most hopeful answer, here, builds upon the approach known as *virtue ethics*. Before we investigate this in depth, however, it's important to consider two other major schools of meta-ethical thought in Western philosophy – *deontological* and *utilitarian* ethics – and the ways in which their offerings are significant but, if taken in isolation, inadequate.

Deontological ethics is interested in questions of moral duty, and the rules of right action that might define such duty. Perhaps the most famous of these is Immanuel Kant's categorical imperative: the argument that each individual should ask of their actions, 'Is the principle upon which I am acting one that should also govern the actions of all other people?'[10] An action is only right, in these terms, if it flows from a moral rule that any right-thinking person would wish to be universal.

Kant's imperative offers a powerful riposte to the prospect of people picking and choosing personal definitions of right and wrong, as well as to the view that no universal ethical standards can be asserted on the basis of experience. As Vallor points out, however, its very universality can render it curiously impotent in the face of present uncertainties.

> Consider the dutiful Kantian today, who must ask herself whether she can will a future in which all our actions are recorded by pervasive surveillance tools, or a future where we all share our lives with social robots . . . How can any of these possible worlds be envisioned with enough clarity to inform a person's will? To envision a world of pervasive and constant surveillance, you need to

know what will be done with the recordings, who might control them, and how they would be accessed or shared . . . [11]

For Vallor, the contingent questions begged by technosocial opacity render the formulation of universal duties incoherent. Unless, of course, we're willing to embrace the uncertainty that deontological ethics seeks to dispense with: to frame the future's duties in terms of what we may owe to one another in specific instances.[12]

There is more to deontology than this, of course – and its emphasis on inalienable duties offers a vital corrective to the tendency to treat human beings as means rather than ends. In particular, philosophers like Luciano Floridi have in recent decades emphasized the ways in which information environments can *themselves* be treated as sites of ethical activity, offering a version of deontology attractively akin to ecology.[13] Yet it is difficult for such maxims to speak to the present with sufficient precision to offer sure guidance. So far as technology is concerned, moreover, the clarity of ethical abstractions becomes less and less clear-cut as they start to be put into practice. As a September 2019 paper in *Nature Machine Intelligence* put it:

> despite an apparent agreement that AI should be 'ethical', there is debate about both what constitutes 'ethical AI' and which ethical requirements, technical standards and best practices are needed for its realization . . . Our results reveal a global convergence emerging around five ethical principles (transparency, justice and fairness, non-maleficence, responsibility and privacy) [but] substantive divergence in relation to how these principles are interpreted, why they are deemed important, what issue, domain or actors they pertain to, and how they should be implemented.[14]

Most 'ethical codes' for systems as complex as AI are much less like computer code than their creators might wish. They are not so much sets of instructions as aspirations, couched in terms that beg more questions than they answer. And the moment any such code starts to be treated as a recipe for inherently ethical machines – as a solution to a known problem, rather than an attempt at diagnosis – it risks becoming at best a category error, and at worst a culpable act of distraction and evasion.

The other major meta-ethical school, *utilitarianism*, founders in a different sense on opacity. Utilitarianism – and the broader ethical category to which it belongs, consequentialism – is based upon the powerfully pragmatic principle that right actions are those aligned with the best possible outcome for the greatest possible number of people (or, depending upon your preferred emphasis, sentient creatures). This approach can also be framed in terms of harm and risk reduction. Right actions are those that do most to reduce preventable suffering, and/or which make catastrophic future events less likely.

While deontological ethics is interested primarily in questions of duty, and thus how intentions map onto generalizable moral rules, utilitarian ethics is interested in the achievement of particular worldly states of affairs. To paraphrase one of the most famous arguments from the philosopher Peter Singer's 2009 book *The Life You Can Save*, almost anyone would naturally leap into a shallow pond in which a child was drowning if the only cost was replacing their new trainers afterwards. Yet, for less than the price of such a pair of trainers, almost everyone living in some degree of comfort can transform the lives of several people suffering elsewhere by, say, donating to a charity like the Against Malaria Foundation. Thus, everyone should either do so, or seek to undertake similarly impactful actions.[15]

A deep tension within utilitarianism is that arguments such as Singer's are simultaneously of immense ethical significance and incomplete. They offer a pragmatic guide to maximizing positive outcomes (and/or reducing negative outcomes) from certain resources, an approach that's directly applicable to countless challenges. Yet at no stage do they constitute a systematic account of human ethical relations. Once we have agreed that certain outcomes are desirable, the reasoned calculus of maximizing these outcomes is precious. But the ethical reasoning supporting such a calculus must inexorably have taken place elsewhere, in contexts within which even an appeal as seemingly self-evident as that of reducing suffering cannot offer clear guidance.

Where, for instance, are the non-subjective moral sentiments to which we might appeal when searching for some 'impersonal' perspective upon which to ground our assessment? As the philosopher Bernard Williams argued in a famous 1977 book, written in dialogue with the philosopher J. J. C. Smart, *Utilitarianism For and Against*, one troubling aspect of utilitarian efforts to come up with a comprehensive account of morality is the 'great simple-mindedness' inherent in its treatment of every output and possibility as ultimately comparable.

> This [simple-mindedness] is not at all the same thing as lack of intellectual sophistication: utilitarianism, both in theory and practice, is alarmingly good at combining technical complexity with simple-mindedness . . . Simple-mindedness consists in having too few thoughts and feelings to match the world as it really is. In private life and the field of personal morality it is often possible to survive in that state . . . But the demands of political reality and the complexities of political thought are obstinately what they are, and in face of them the simple-mindedness of utilitarianism disqualifies it totally. The important issues that utilitarianism raises should be discussed in contexts more rewarding than that of utilitarianism itself.[16]

At the other end of the scale to Singer's focus on immediately preventable suffering – a divergence that itself suggests the difficulty of reconciling rival utilitarian framings – so-called 'longtermist' philosophers such as Toby Ord and William MacAskill (building on Derek Parfit's foundational work) have in recent years emphasized the importance of consequentialist caveats aimed at avoiding civilizational disaster.[17] Lists of these existential threats typically include nuclear war, global epidemics, meteorite strikes and climate breakdown, not to mention hostile machine superintelligences.[18]

It would undoubtedly be wise for societies to spend more time thinking about these things – and MacAskill sketches an appealing approach based upon keeping civilization's options open. Yet there can also be something self-cancelling about so relentless a focus on distant futures: an instrumental attitude towards present injustice, inequity and complexity that risks sabotaging the legitimacy of even the most admirable aims. Despite the clarity with which it's possible to conjure prospects of annihilation, moreover, a world of predictable systems whose impacts can be fine-tuned by impartial experts does not and cannot exist. All technological systems are embedded within human, political and economic contexts – and engaging with these will *always* entail incommensurable assumptions, ongoing negotiations and incalculable outcomes. Kate Crawford explains this eloquently in her 2021 book *Atlas of AI*.

> Artificial intelligence is both embodied and material, made from natural resources, fuel, human labor, infrastructures, logistics, histories, and classifications. AI systems are not autonomous, rational, or able to discern anything without extensive computationally intensive training . . . In fact, artificial intelligence as we know it depends entirely on a much wider set of political and social structures.[19]

Ultimately, despite the admirable significance of utilitarian analyses within certain domains, there is no one great test to be passed, no single societal consensus or paradigm to be shifted – and no way of imposing alleged solutions into such spaces without silencing many of those voices that most need to be heard. Rather, there is the unfolding collective challenge of finding ways of flourishing under conditions of technosocial opacity; and, incrementally, imperfectly, of creating *virtuous cycles* of technology's development, interrogation and deployment.

The power of virtuous processes

A central contention of virtue ethics is that, given the profound uncertainties surrounding each unfolding life, no one trajectory is guaranteed to provide purpose or contentment; but that it is possible to describe the kind of practices and aptitudes compatible with the fulfilment of human potential. Such fulfilment is termed, in the Aristotelian virtue ethical tradition, *eudaimonia*. What does *eudaimonia* entail? The philosopher and classicist Edith Hall teases out some of its complexities in her 2018 book *Aristotle's Way*.

> The eu- prefix (pronounced like 'you') means 'well' or 'good'; the daimonia element comes from a word with a whole range of meanings – divine being, divine power, guardian spirit, fortune, or lot in life. So eudaimonia came to mean well-being or prosperity, which certainly includes contentment. But it is far more active than 'contentment'. You 'do' eudaimonia; it requires positive input. In fact, for Aristotle, happiness is activity (praxis). He points out that if it were an emotional disposition which some people are either born with or not, then it could be possessed by a man who spent his life asleep, 'living the life of a vegetable . . .'[20]

Aristotle is, Hall notes, 'usefully gregarious and concrete as a model for virtue in practice' – which isn't the same thing as being timelessly correct. Aristotle was wrong about plenty of things (gender politics and slavery among them). In bequeathing the world a view of ethics that insists upon their concrete, contingent quality, however, he provided a framework well suited for addressing the tensions and interdependencies anatomized so far – not to mention a philosophy compatible with a host of other traditions committed to purposeful self-development, from Thomist Christianity to the works of the Islamic scholar al-Fārābī, the Jewish scholar Maimonides and much of Buddhism, Daoism and Confucianism.

In particular, virtue ethics is committed to the idea that moral character lies at the heart of ethics; and that it is primarily by working on our own character that we become able to treat others well. Moral character is a capacious concept. It relies not on fixed rules of wrong and right action, but rather on practising virtuous behaviours in day-to-day life – and the psychological significance of role models upon behaviour and beliefs. Every action, no matter how small, is potentially a precedent. Similarly, inactions and happenstance are of great significance. To be disadvantaged, abused or unfortunate is to be confronted by obstacles to thriving that it may prove impossible to overcome. In this sense, civic virtues such as respect for justice, fairness and liberty – and the communal cultivation of these – can be of greater weight than purely personal achievements.

Above all, virtue ethics is determinedly modest in its ambitions. It sees thriving and goodness as lifelong journeys with no final destination, and even the best of us as only too human. As the philosopher Julian Baggini put it in his 2020 book *The Godless Gospel*, an exploration of Jesus' life as a model for secular ethics:

> One neglected feature of Jesus's example is that he models the need for work on the self. The supposed divinity of Christ tends to make

us think of his goodness as being inherent, but this is not how he is portrayed in the Gospels . . . Even someone as morally gifted as Jesus needed time for his wisdom to grow, and that wisdom needed constant nurturing.[21]

At this point, it's useful to consider a concrete example of virtue in practice when it comes to tech; and, in particular, of what it may mean to align a technology's development and deployment with the freedom and empowerment of those associated with it.

In November 2016, the researcher Joy Buolamwini – then a grad student at MIT – spoke at TEDxBeaconStreet about facial recognition systems and race. When she was an undergraduate at Georgia Tech studying computer science, Buolamwini explained, she used to work on so-called social robots – and soon discovered that the robot she was using couldn't 'see' her because of the colour of her skin. In a pre-emption of the problem with some proctoring software discussed in chapter seven, she found that she had to 'borrow' her lighter-skinned roommate's face in order to complete a project. Soon after this, she visited Hong Kong to take part in an entrepreneurship competition and paid a visit to a local start-up that was demonstrating one of its social robots. 'You can probably guess', she said, what happened next.

> The demo worked on everybody until it got to me . . . It couldn't detect my face. I asked the developers what was going on, and it turned out we had used the same generic facial recognition software. Halfway around the world, I learned that algorithmic bias can travel as quickly as it takes to download some files off of the internet.[22]

As a steady stream of examples has subsequently emphasized – from Zoom calls 'cutting off' the heads of those with dark skin to

Twitter algorithms automatically placing white faces at the centre of cropped images[23] – Buolamwini was being excluded by default from such categories as 'normal', 'significant' and even 'human'. Importantly, however, she was also far from a passive victim.

In order to 'see' anything, a machine-learning algorithm must be trained by exposing it to samples of whatever it is supposed to recognize: in this case, hundreds of thousands of examples of both faces and things-that-are-not-faces. If only certain types of faces are included in the training set, those who deviate too far from their norm will be harder to detect. But all of this, Buolamwini notes, embodies not so much the implacable verdict of an automated system as the explicit product of a series of human choices.

> Training sets don't just materialise out of nowhere. We actually can create them. So there's an opportunity to create full-spectrum training sets that reflect a richer portrait of humanity . . . we can start thinking about how we create more inclusive code and employ inclusive coding practices. It really starts with people. So *who* codes matters. Are we creating full-spectrum teams with diverse individuals who can check each other's blind spots? On the technical side, *how* we code matters. Are we factoring in fairness as we're developing systems? And finally, *why* we code matters. We've used tools of computational creation to unlock immense wealth. We now have the opportunity to unlock even greater equality if we make social change a priority and not an afterthought.[24]

Why, how, who? For all the complexities of the answers they demand, the questions that unlock the black box of encoded injustice couldn't be simpler. And this in turn suggests some of the most fundamental things we can say about the biases, prejudices and distortions latent in so many systems: that all of these are only ever invisible to *somebody*, *somewhere*; and that it's only a narrowly deterministic narrative that allows this *somebody* to plead ignorance on behalf of humanity as a whole. All too often, the issue that 'we' couldn't have been expected to anticipate turns out to have been both predicted and protested by plenty of people only too familiar with its details – but whose voices and experiences weren't welcome in the rooms where it was encoded.[25]

It is seven years, at the time of writing, since Buolamwini's talk. She has since founded a movement known as the Algorithmic Justice League and helped build change, alongside like-minded others, in the direction of equitable and accountable AI.[26] Yet the very flaw she identified continues to create divisions and disadvantages, as do countless other inequities and exclusions. What is to be done? The answer is as much about the people and priorities present (and absent) in boardrooms and workplaces as it is about data or code; and it points towards the heart of the problem for tech ethics itself. When it comes to technology, it's not enough that we seek either virtuous tools or virtuous people. Rather, we need to ask what it means for the ongoing process of designing, debating and deploying a technology *itself* to be a virtuous one.

Acknowledging our mutual dependency

If prejudice and injustice are inscribed in the data we feed into machines, then scrutinizing this data presents a profound ethical opportunity: a chance simultaneously to recognize and redress structural inequalities, exclusions and injustices. Importantly, however, it will never be adequate to focus only (or even primarily) upon data itself. As the researchers Alex Hanna, Emily Denton, Razvan Amironesei, Andrew Smart and Hilary Nicole argued in a December 2020 essay for *Logic(s)* magazine, there is a gaping absence at the heart of any argument that ethical issues can be resolved solely by relying on big companies to build up bigger and better datasets.

A particularly pernicious consequence of focusing solely on data is that discussions of the 'fairness' of AI systems become merely about having sufficient data. When failures are attributed to the underrepresentation of a marginalized population within a dataset, solutions are subsumed to a logic of accumulation; the underlying presumption being that larger and more diverse datasets will eventually morph into (mythical) unbiased datasets. According to this view, firms that already sit on massive caches of data and computing power – large tech companies and AI-centric startups – are the only ones that can make models more 'fair.'[27]

To return to this chapter's subtitle, we cannot afford to believe that some divine dataset will ever dissolve all difficulties – or to misread size, speed and complexity as their own justifications. Much as there's no perfectly impartial calculus according to which a utilitarian can weigh the world in their scales, so appeals to 'unbiased' datasets invoke a set of conditions that can never be met: a world in which no value-laden choices or preferences exist around a technology's research, development, deployment, governance and regulation. Consequentialist attempts to maximize certain fair or desirable outcomes are a tremendously powerful tool. But they are not, and cannot be, the full story. By contrast, to think about technology's development and deployment in terms of virtue is to embrace its entwining with our minds and natures – while refusing the pretence that either it or we can ever be perfected.

It's at this point that several of this book's themes come together: the significance of multiplicity, uncertainty and mutual attentiveness; the obligation to update old assumptions in the light of new knowledge; an alertness to the values embedded in information environments themselves. One gift of virtue ethics is that it requires us to address this context through the lens of each life's potentials and dignity: that we acknowledge the weighty demands made of us by hopes of growth and thriving. And perhaps the weightiest demand of all is the one from which all the rest must flow: that we acknowledge the depths of our fallibility, vulnerability and mutual dependency.

In his 1999 book *Dependent Rational Animals* – an expansion of a series of lectures delivered to the American Philosophical Association in 1997, sixteen years after the publication of *After Virtue* – the philosopher Alasdair MacIntyre makes the case that discussing human existence in terms of the 'normal' capabilities of healthy, seemingly autonomous adults is itself a profound ethical category error. This is not only because to do so is to ignore the arbitrariness of the world's inequalities, but also because our existence is defined in the most fundamental sense by dependency: by our species' extended infancy and childhood; by sickness, infirmity, and age; by the collective nature of culture, trade and technology. If we are meaningfully to discuss life as it is actually lived, MacIntyre suggests, we must begin not with a snapshot of some notionally independent adult, but rather with the demand that:

> those who are no longer children recognise in children what they once were, that those who are not yet disabled by age recognise in the old what they are moving towards becoming, and that those who are not ill or injured recognise in the ill and injured what they often have been and will be and always may be.[28]

It also matters, MacIntyre continues, that this recognition of mutual dependency is couched in terms of our bodies as much as of our minds. 'I now judge that I was in error,' he writes in the preface, 'in supposing an ethics independent of biology to be possible,'[29] a reference to his previous stance in *After Virtue*. Virtue is embodiment, not abstraction. It is inscribed upon the vulnerability of mortal flesh. To be human is to be born into utter helplessness, in circumstances beyond our choosing. It is to grow and change, constrained by these circumstances and biological inheritance. It is to achieve some measure of independence, for a time, in the context of our planet's and our societies' vast networks of exchange and competition. And it is to seek not only survival but also – so long as the body's basic needs are met – some form of flourishing or contentment. There is no final victory, no guarantee of success, and no infallible guidance. There is only the contingent business of trying, together, to live and to know ourselves a little better.

All of the above entails, to return to a phrase from chapter seven, moral labour whose difficulty and significance are inextricably linked. As the parent of two young children, one of the hardest lessons I have struggled to master – like most parents – is the fact that my children's desires are an imperfect guide to their wellbeing. Making their lives easier is not always the best way to prepare them for life. Much like the students Frischmann and Selinger describe in *Re-Engineering Humanity*, it's more important for me gradually to help them develop a measure of self-control, fairness and ambition – and to show them how trust can be earned – than it is for me constantly to monitor and intervene in everything they do.

Also like many parents, a second lesson I'm still trying to learn is that the person who all too often needs to improve their self-control is *me*. To love and to nurture other human beings brings pain as well as joy; frustration and exhaustion as well as delight; the prospect of devastating loss alongside the gains of consuming love. And these satisfactions and sacrifices can't tidily be separated.

To withdraw your care from any relationship is to make yourself less vulnerable, for a price: it's to diminish what you risk and give, but also what you can receive and gain. As the philosopher Virginia Held puts it in her 1997 book *The Ethics of Care* – a key text for the subset of virtue ethics known as care ethics – 'without some level of caring concern for other human beings, we cannot have any morality.'[30] Among all the virtues it's possible to enumerate, it is loving care that most closely binds us to the world and one another; that connects the universal with the most intimate facts of existence.

I could make my life easier by outsourcing my children's education, discipline and nurture to the nudges of expert systems, much as a government might choose to reward or punish its citizens' actions via ubiquitous surveillance. In each case, however, the fantasy of an optimized existence is one that hollows out not only people's relations with each other, but also the value of most other things worth pursuing. It seeks to impose an empty vision of perfectibility in place of the purposeful, mutual struggles through which human dignity and potential are asserted and sustained.

Love and dissent

'Why does truly trying to know the world we live in, the history that makes us, matter?' asks the author Elaine Castillo in her incandescent 2022 essay collection *How to Read Now*. The answer, she suggests, is love.

> Loving this world, loving being alive in it, means living up to that world; living up to that love. I can't say I love this world or living in it if I don't bother to know it; indeed, be known by it.[31]

To love is to attend to the world, the self, the other; to try to read these without erasure or denial; to embrace change, contradiction and difference rather than reaching for comforting simplicities. Similarly, it is only when those designing and deploying a technology start seeking out others' experience rather than making assumptions on their behalf – the moment they start embodying open questions like *why*, *how* and *who* in a design process rather than declaring their own preferences to be synonymous with the 'logic' of technology – that they begin, for the first time, to see technology as it actually is. That is, they begin to see the human-made world as one that its creators and maintainers both bear responsibility for and are constantly instantiating this responsibility within.

As the twenty-first century unfolds, it is becoming all too easy for ethical outsourcing, behavioural engineering and surveillance to infiltrate ever further into our lives – and to do so in the name of maintaining standards, preventing deceit and providing support. Yet such claims are hollow at the core: not because they are ineffective (after all, it's their putative effectiveness and efficiency that makes them seductive), but because they are too often corrosive of the very possibility of earning or bestowing trust; of the private spaces within which self-knowledge, self-authorship and rich mutual engagement can occur.

Against this, what's needed is an explicitly ethical understanding of the assumptions embodied in technologies' design and deployment: one alive to modernity's complexity, opacity and interdependencies. Neither expert condescension nor the decontextualized praise of personal responsibility are adequate for such tasks. In the meeting place between virtue, principle and pragmatism, however, there is something sufficiently modest and humane to speak to the grandest of timescales: that understands the practices flourishing can arise from, and its dependency upon systems that enhance rather than diminish our faith in one another.

Central to the idea of virtue is its practical cultivation over the course of each life and, in parallel with this, a belief in the human potential to grow beyond our beginnings: to follow role models and potentially to become one. If much of this sounds abstract, its implications – as befits a philosophical tradition emphasizing the importance of *praxis* (thoughtful action) and *phronesis* (practical wisdom) – are only too tangible. What should citizenship look like in a digital age? How can we dismantle the injustices embedded in some conceptions of 'normality' and desirability, or push back against the diminishment that comes from the delegation of education, work and governance to opaque, unchallengeable systems? The answer, in each case, begins with the rejection of wishful delusions – and the insistence that it is *our* needs and capabilities, not the hypothetical beneficence of superhuman systems, that define a future worth believing in.

Many of our most urgent opportunities for action require us to push back against narratives of inevitability and optimization. From facial recognition systems to the normalization of surveillance, from autonomous weapons to weaponized social media ecosystems, there has never been a stronger case for mindful delay, dissent and disavowal – and for forms of ethical thinking that place such dissent upon firm foundations. As the philosopher Carissa Véliz puts it in her 2020 book *Privacy is Power*, to speak of virtue and lived experience in present times is necessarily to speak of righteous anger as well as cool consideration; of the fact that human growth and flourishing are sometimes best served by resistance.

Aristotle argued that part of what being virtuous is all about is having emotions that are appropriate to the circumstances. When your right to privacy is violated, it is appropriate to feel moral indignation. It is not appropriate to feel indifference or resignation. Do not submit to injustice. Do not think yourself powerless – you're not.[32]

There is always a choice. My hope is that, together, we can more often make it a wise one.

CHAPTER 12

Death and life:
The delusion of perpetual progress

Over two thousand years ago, Plato cautioned us that writing divides words from their speakers: that it creates texts that cannot speak for themselves, and may thus be misunderstood and abused. He couldn't have anticipated all that writing, literacy and information technology would bring. But he knew that the trajectory of such developments signalled profound change – and that this change mattered precisely because human minds and communities could so readily embrace and inhabit it.

Our capacity for change is the most deeply rooted of human gifts. From generation to generation, we teach, learn and alter ourselves. And we do so with neither a final destination nor any ultimate set of answers in hand. We are self-made and self-remaking: inheritors and custodians of a contingent world.

Sitting at my desk in a warm, dry room, drinking coffee, typing words onto a screen, my life is a grateful compendium of technosocial enhancements. Artificial light and heat illuminate and warm me. Artificial shelter protects me. Clothes keep me warm, shoes cushion my feet, glasses improve my eyesight. Several artificial teeth sit in my mouth, while my blood contains the traces of technologically-induced immunity against a host of diseases. I am father to a son who would probably have died in childbirth, along with his mother, a few hundred years ago. My phone pulses in my pocket with a message from friends halfway across the world. My home is stocked with foods others have packaged and prepared for my consumption, as well as unlimited supplies of clean water and power. I am privileged, yet far from unique. And I believe that a better world is one in which more people enjoy the most precious things I possess: freedom from want and fear; security of body and mind; a chance to pursue their versions of a life worth living.

A hundred years from now, what else might my children's grandchildren add to the lists above? Genes engineered to protect them from a gamut of harms? Information technologies integrated into their bodies and brains as well as their minds? Machine companions tending to their every need? All this may come to pass. Even if it does, however, it will remain part of the same human story I and my peers are living. And it will be the story of social animals for whom freedom, security, hope and love remain the most significant affordances a society can offer – not to mention the existence of a planet capable of sustaining us and other life in the first place.

This in turn suggests the central flaw of so many projects of technological optimization and self-transformation: the shallowness of their relationship with history and time. Self-enhancement is neither foolish nor impossible. Its limits are indeed unknowable, its potentials endless. But none of this moves us beyond our humanity any more than it circumvents our conflicts, needs and ambivalences – or our continuity with other species. We can neither wish, reason nor invent our way out of every present challenge, not least because each undertaking breeds new risks and tensions: revelations that must, in turn, be acknowledged and accommodated.

Similarly, we can neither create nor become perfect machines. What *is* within our scope is to constrain the self-deceptions, avarice and cynicism that attend our existence. But we can only hope to do so if we're able to look *within* and *around* us with compassion as well as wonder; and if we can escape the delusion that machine-made salvation will ever be at hand.

The difference between actual and imagined progress

If you're looking for a contemporary vision of machine-made salvation, the cryopreservation industry is one contender. Here's how the website of the Californian cryonics company Alcor describes its business:

> Cryonics is the practice of preserving life by pausing the dying process using subfreezing temperatures with the intent of restoring good health with medical technology in the future.[1]

And here is its explanation of what this might mean for *you*:

> We're making cryonics accessible to everyone. With low monthly dues and an insurance policy, you're all set. When the time comes, we'll perform your cryopreservation at our state-of-the-art facilities. Patients are kept in secure, long-term cryogenic dewars [vacuum flasks] until revival. Welcome to your future.[2]

While cryonics is rooted (albeit distantly) in proven research, what's intriguing about enterprises like Alcor is as much about narrative framing as scientific possibility.[3]

Consider the ambition to 'pause' the 'dying process', with its implicit echo of a movie or videogame: an entertainment that can be stopped and started at the touch of a button. The promise of future technology here isn't just a punt on a distant prospect. It's also a story about how life itself should be valued and understood. In order accurately to define death, the Alcor website suggests that we need to see it not as a physical but rather as an *informational* concept.

> Death is only permanent when the structures encoding memory and personality (necessary for consciousness) have become so disrupted that it becomes theoretically impossible to recover the person. This is called 'information-theoretic death'. Any other definition of death is arbitrary, and subject to revision.[4]

The only 'non-arbitrary' definition of death, according to this line of argument, concerns the informational state representing a particular person. This definition in turn relies upon the assumption that people are, in theory, synonymous with certain informational states; and that, in practice, nobody can truly be considered dead until it is impossible to 'recover' this state.

All of this connects to further speculations that lie far outside the social and technological mainstream, but that are staples of *transhumanist* discourse: the branch of futurology devoted to humanity's capacity to augment and transcend itself through technology.[5] What does it mean to suggest that a person can be expressed in informational form; or that sufficiently advanced technologies will be able to read or replicate this state? Will the revival of frozen brains eventually prove feasible, or human informational states be uploaded into synthesized bodies or simulated environments?

Going 'beyond the human' here offers an exultant antidote to imaginative constraint: a series of thought experiments inviting us to test foundational assumptions about life, death, consciousness and reality. Yet at the same time, it shrinks the scope of what is at stake. If, sooner or later, everyone and everything is destined to become a series of data points read by machines, it's tempting to abandon other forms of custodianship and caring in the meantime – and to treat the preservation of information as your only ethical imperative.

A great deal has been written about such speculations, up to and including the theory that our entire existence is in fact contained within a computer simulation.[6] While these purport to be plausible analyses of possible futures, however, what's most noteworthy about ideas like information-theoretic death and the so-called 'simulation hypothesis' is the common ground they share with a mindset they nominally oppose: that of faith.

This is because they are fundamentally *metaphysical* forms of speculation. Unlike empirical science – or the pragmatic business of enhancing human minds and bodies through existing technologies – they can neither be verified nor disproved by any investigation of material reality. Instead, they posit the extension of present trends towards a hypothetical tipping point, then reason back from this to justify particular actions and attitudes: not least, that the imagined capabilities of a distant future may be able to redeem all present wrongs and losses. We are back in the familiar terrain of the Singularity: a place distinguished by the gulf between the *actual* dilemmas that have immemorially attended technological development and the *imagined* dilemmas that attend projections made on its basis.

As for those who do believe we are moving beyond the human, what do they think we are moving *towards*? One pleasingly literal suggestion is that transhumanism will necessarily lead to a *post-human* state: an 'after-human' future marking a disjunction between past and present so radical that future generations will no longer be human in the sense we currently are. As the Transhumanist FAQ website explains:

> It is sometimes useful to talk about possible future beings whose basic capacities so radically exceed those of present humans as to be no longer unambiguously human by our current standards. The standard word for such beings is 'posthuman.'[7]

In one sense, posthumanism invokes the entirely reasonable point that we are constantly discarding older visions of ourselves: that our cultural and technological capabilities are a form of ceaseless collective augmentation. In another sense, however, its explicitly technological vision of rupture – and its relegation of present (let alone past) humans to the level of supplicants at the feet of those yet to come – is an avowedly ideological form of speculation. As the philosopher John Gray points out in his 2011 book *The Immortalization Commission*, such a story is in effect an inversion of the most ancient myth of them all, that of creation.

> In this version of theism it is not God that creates humans. Rather, humans are God in the making . . . [8]

At stake, Gray suggests, is a philosophy of progress whose alleged secularism disguises not only an essentially religious impulse but also an explicitly religious genealogy: a direct line of descent from so-called 'process theologies', which emphasize the unfolding of a divine destiny or purpose over time.

> [In] process theology . . . a number of twentieth-century theologians, mostly American, imagined God emerging from within the human world. Rather than an eternal reality, God was seen as the end-point of evolution . . . Process theology is one more philosophy of progress – an attempt to solve the problem of evil by positing its disappearance over time . . . Meliorism – the belief that human life can be gradually improved – is usually seen as a secular world-view. But the idea of progress originates in religion, in the view of history as story of redemption from evil. Philosophies of progress are secular religions of salvation in time, and so, too, is the Singularity.[9]

Are all invocations of 'progress' delusional? As this chapter's subtitle suggests, the hope that technology can gift humanity a perpetual form of progress has little to do with reality. But this doesn't mean we should ignore the countless ways in which human societies and lives *have* meaningfully improved over the millennia — or what it means to continue extending and defending these developments. From health and education to cultural and scientific knowledge, from fundamental rights and equalities to material comforts, our species has transformed the scope and prospects of human life beyond recognition. Such achievements are ongoing and contested; but no less real because of this. Indeed, it's the very fact that progress in all of these areas is at once particular and imperfect that distinguishes it from the abstract faith Gray criticizes.

Just like real lives, *real* improvements to these lives will always be incremental, contestable and contingent. Diseases treated, vaccines developed, research conducted; books written, systems designed, software coded; art produced, evidence gathered, arguments advanced; voices heard, losses comforted, injustices acknowledged: progress has been made across these and other domains not because it is an inexorable by-product of innovation or history, but because countless collective battles have been — gradually, imperfectly, unevenly — fought and won.

None of this was preordained and none of it is guaranteed to last. Every advance humanity has made demands constant maintenance and defence; while that which is yet to be won demands argument and advocacy. Like a virtuous life — like any existence worth sharing or bequeathing — meaningful progress is a constellation of acts and insights. It is a fine and necessary thing for us to have faith in one another. But the last thing we can afford is to project this faith onto time, technology and history. Similarly, scientific research can ill afford to confuse dreams of certainty with the underlying nature of reality. As Gray notes:

The most rigorous investigation reveals a world riddled with chaos in which human will is finally powerless. All things may be possible, but not for us. This is not a conclusion many people are ready to accept. There is a persisting need to believe that the order that is supposed to exist in the human mind reflects one that exists in the world. A contrary view seems more plausible: the more pleasing any view of things is to the human mind the less likely it is to reflect reality.[10]

We're coming close, here, to something that lies beneath *all* the delusions this book has anatomized: our faith that the future is something we can know, as if it were an island on the horizon and technology a telescope; our eager over-confidence that we are voyaging towards secular salvation. Ours is indeed an exponential age. Yet, in evolutionary terms, even the most radical changes are transitions between different kinds of stability. And this means that our defining challenge is not so much to ride a steepening curve towards transcendence as to negotiate the terms of a new equilibrium.

Finding purpose through those who come after us

In a 2010 installation of the artist Tino Sehgal's 2006 work *This Progress* in New York, visitors walked into the bottom of the Guggenheim museum's spiral rotunda to be met by a child of between seven and eleven years old who engaged them in conversation. 'This is a work by Tino Sehgal. What is progress?' the child asked, inviting the visitor to walk with them. As the conversation unfolded they ascended, until the visitor was handed over to their next 'interpreter': a teenager, who continued to walk and to talk. Next came an adult, then someone in old age. Then it was over.

The central question at stake was simple enough. But it's a safe bet that it received almost as many answers as there were conversations. The point of *This Progress* was not so much any individual answer as the experience itself: a journey, a series of dialogues, an opportunity to reflect upon time and change. As the description for the Guggenheim's online collection puts it:

> Because of the individualized conception of Sehgal's work, experiences of each piece are highly subjective and might change dramatically on repeat visits, leading to discussions about a wide range of topics such as sustainability, economics, politics, social reform, or personal growth. These varied results from a single scenario are the crux of Sehgal's singularly elastic and provocative practice. By setting the stage for a conversation, rather than opting for more traditional one-way spectatorship, Sehgal puts the spectator and the work on equal footing . . . [11]

Dialogue is the key. And the prize it offers – as Plato knew – is not so much certainty as a more richly inhabited relationship with enduring questions. What do *you* mean when you speak of progress, purpose, value or justice? Why? If you're not sure how to answer, a good place to start is to try explaining yourself to someone else: to exchange ideas in good faith. Like life itself, arriving at the end of a dialogue isn't what makes the experience worthwhile. The journey – what you bring and give along the way – is the place where meanings must be made.

Sehgal's *This Progress* puts its questions into the mouths of young and old, suggesting both the shifting perspective of all our lives and their common thread. What does it mean, though, for these conversations to take place across countless lives: to be handed down from generation to generation? In his 2013 book *Death and the Afterlife*, the philosopher Samuel Scheffler proposes a thought experiment that dramatizes this question.

> Suppose you knew that, although you yourself would live a normal life span, the earth would be completely destroyed thirty days after your death in a collision with a giant asteroid. How would this knowledge affect your attitudes during the remainder of your life?[12]

My reaction to Scheffler's story is typical, in that the idea of this world ceasing to exist fills me with despair. This may seem fair enough. Yet, Scheffler points out, it is in some ways unreasonable. How can my own life be damaged by events that, by definition, I will not and cannot experience? Why, moreover, should the notion of imminent planetary destruction matter so much given that – in the long term – I fully accept that all things must pass, the Earth and my own species included?

One possible answer is that neither logic nor narcissism loom as large within my world-view as I might think; and that, even though I can only encounter reality through the lens of my own experiences, the nature of my existence is dependent upon larger continuities. Indeed, Scheffler points out, it may be the case that the significance of *everything* I care about is bound up with some belief in its continuity. The people and things I love; the places, the poems, the creatures; the pleasures of food, drink and laughter. If I knew all these would be annihilated alongside me I would, I suspect, spend every day of my life mourning them.

It's a commonplace that most people prize the continuity of *some* things over their own existence. Like most parents, I would sacrifice my life if it meant saving those of my children. Yet my children will undoubtedly die, just as all things must pass. So why should *any* entity – a child, a species, an idea – command greater loyalty than the only life I possess? As Scheffler notes, these thoughts point towards a central mystery in all of our relationships with time: a sense in which our judgements are inherently relative not only to real or remembered comparators, but also to our hopes for a future we will never know:

> the coming into existence of people we do not know and love matters more to us than our own survival and the survival of the people we do know and love . . . This is a remarkable fact which should get more attention than it does in thinking about the nature and limits of our personal egoism.[13]

There is certainly something remarkable about the weight of the future within present lives. Yet is there really a mystery here? Only if you begin by assuming that egoism is an unimpeachably rational perspective – or that it's irrational to found values and purposes upon hope. Viewed through the lens of an individual life, the supremacy of personal experience may seem self-evident. But from the species perspective, selflessness is our greatest gift. To care for those who come after you; to love the grand continuity of culture, technology and learning; to despise whatever denies the yet-to-be-born their chance of thriving: all of these are exquisitely adaptive responses to time and uncertainty. And what counts most of all is *continuity* itself: not so much any particular vision of progress as faith that others' thriving will remain possible. It is the unknown and unknowable nature of our descendants that defines the strangeness of this hope – and that gives the lie to fantasies of machine-engineered exceptionalism.

Another way of putting this is that while the future is where all our hopes must reside, dreams of optimization only make sense if you shrink life to a predictable game. Evolution, the universe and everything are often presented as fond of efficiency: the fittest survive, the unfit demise, humanity ascends its alp of achievement. Yet something very different should surely be read into the mad, glorious variety of organisms crawling, striding, flapping, gliding and inching their way across our Earth. Life seeks to beget other life, and this superabundance has over the last few billion years proved an excellent survival strategy. It's only the sheer variety, depth and tenacity of life's colonization of our planet that has seen it continue through periodic waves of mass extinction – and flourish into equilibria of fractal complexity. Life as a whole is both supremely adaptive and stupendously arbitrary. It has to be, because the universal background against which it exists is as diametrically opposed to any linear definition of progress as you could imagine: a cosmic unravelling known as entropy, its logic a downhill ride all the way from time's first moment.

Redundancy, superfluity, excess; compassion, collaboration, nurture: *these* are the grand strategies of endurance across deep time. These are the roots of resilience, because they do not pretend to know or control all that is to come, let alone to be capable of optimizing our species towards one century's dreams of destiny. And the same is true of our mental lives – not to mention the tools we create, and through which we are in turn remade. Efficiency is a fine thing when it allows us to leap further in our aspirations while treading more lightly. But it mustn't be made an end in itself or confused with salvation. By narrowing our focus – by cleaving away complexity – we make ourselves fragile in the face of time.

The journey is not only more than a means. It's also the place where any ends whatsoever must exist. If life represents an extraordinary species of organized resistance to disorder, it is also grandly, definingly inefficient. And if we humans are supremely gifted at anything, it is at achieving a similar superabundance: at loving, hoping and striving afresh; at grasping towards continuities in sure knowledge of our transience.

Depending upon your perspective, we are either an evolutionary miracle or an ongoing crisis for everything else living on this planet – or, perhaps, a bewildered mix of both. In each case, we are not to be explained away by final destinations or speed of transit. Up the spiral we tread, ascending through our years, telling our stories. A thousand different answers to a question in the mouths of young and old, asked for its own sake, accumulating increments of understanding; a tale told ten thousand times: these are richer and more hopeful propositions for our species than dreams of self-perfection.

Gathering a dozen delusions

In his 2020 book *The Good Ancestor*, the philosopher and author Roman Krznaric asks what it means to bequeath our descendants a world worth living in. The answer, he suggests, is bound up with a 'tug of war for time' in which short-term pressures compete with long-term thinking for a critical mass of human belief.

What does long-term thinking entail? For Krznaric, we need to combine humility in the face of deep time with ambition in the face of its challenges. We need 'cathedral thinking' to help us plan projects beyond a human lifetime; we need 'intergenerational justice' to reframe fairness and equality as a contract between multiple generations; we need 'holistic forecasting' to help us envision multiple possible pathways for civilization. Above all, we need to focus upon a goal that is both transcendent and achievable: ensuring the long-term thriving of our own and other species 'within the means of earth's crucial life-supporting systems'.[14]

Krznaric is married to the author and economist Kate Raworth, and her 2017 book *Doughnut Economics* sets out a similarly forceful vision for a sustainable future. Arguing that GDP growth is a misguided and overly simplistic goal for governments, Raworth replaces traditional graphic representations of economic activity – in which a single line traces the (hopefully) ever-increasing wealth of a nation over time – with her titular 'doughnut' graph, which represents not growth but equilibrium. Instead of plotting time along an endlessly advancing axis, Raworth's doughnut is a closed loop inviting us to assess humanity's performance across nine ecological and twelve social domains.

What follows if we embrace such measures? In place of a graph's endless unspooling, the doughnut is an instrument attuned to crises and consequences; an oscillating measure of systemic resilience. And with this comes a vision of the future committed as much to preservation and remedy as innovation; one whose measures of human achievement are not numbers indefinitely ticking upwards, but potentials sustained from generation to generation. By separating progress from presumptions of growth, different possibilities come to the fore.

Every graph is also a story – and there are always more stories waiting to be told. Across a dozen chapters, this book has sought to challenge some of the stories the twenty-first century most commonly tells about technology. Contrary to the *delusion of inevitability*, neither the past nor the future can be understood as an inexorable progression of innovation-driven developments – while our species' unique capabilities do not, as the *delusion of mastery* would have it, make us the masters of the forces governing all other life. We are, rather, uniquely vulnerable and co-dependent; remarkable above all in our collective capacities for teaching, learning and changing, both over the course of a life and between countless lives. The claim of the *delusion of brutality*, that conflict and competition are the definitive features of our own and our creations' evolution, similarly cannot survive historical scrutiny.

Tools and technology have been our evolutionary companions since before the beginning. Despite the *'it' and 'us'* delusion, and its claim that technologies are optional and external to our natures, we live and dream outside the confines of our individual bodies:

through artefacts, information and ideas; through the unbounded arena of signs and symbols. All this is possible precisely because we can never access the external world impartially – but are, contrary to the *delusion of literal-mindedness*, in the constant collective business of updating the controlled hallucination of consciousness.

In the face of such complexity, it is natural for us to succumb to the *delusion of comprehension* and resolve events into tidy tales of cause and consequence; into idealized projections of our mental lives. But this ignores many of the most important lessons that time, fate and survival can teach. What's required of us is new forms of humility in the face of our century's interlocking challenges – and an undeceived attentiveness to the hidden histories and unintended consequences that surround even our greatest achievements.

Like its makers and maintainers, the human-made world is suffused with wants and values; with inclinations and assumptions that reshape us even as we shape them. The *delusion of neutrality* is a way of shirking the ethical import of this fact. Contrary to the *delusion of magical thinking*, human society is not a problem awaiting technological solutions, nor innovation a form of religion – while, contrary to the *anthropomorphic delusion*, human minds are as misleading a model for understanding machines as machines are for understanding ourselves.

The exponentially increasing power of automated systems is – if we can only free ourselves from wishful anthropomorphism – one of the most remarkable and hopeful facts of our age. Indeed, the present's greatest riches and opportunities can *only* be accessed through automated systems of ever-increasing brilliance. But it's the gulf between us and the uncomprehending competence of our creations that matters most when it comes to negotiating richer human–machine relationships. Neither perfect machines nor perfectly reasonable decisions can exist in the real world; and denigrating human capabilities in comparison to *delusions of machine perfection* is a sure way of stunting our dignity, autonomy and potential.

Similarly, contrary to the *delusion of divine data*, there is no point at which sufficient quantities of data will ever become self-interpreting or free from presumption – and no sense in which a technocratic elite will ever be able to speak on behalf of the world. In order to grapple fruitfully with these challenges, we need to move beyond dreams of ethical optimization and outsourcing, and to focus instead on designing and deploying technologies through *virtuous processes* – while attending closely to questions of *why*, *how* and *whether* any systems should exist in the first place.

Finally, despite the eagerness with which its prophets preach technological transcendence, machine-assisted salvation is not at hand. We have immemorially enhanced our bodies and minds through technology, and will surely continue to do so. But – contrary to the *delusion of perpetual progress* – this will not move us beyond conflict, fallibility and self-deception, nor transform us into undying machines. Rather, the best measures of our successes and failures will always lie in those inheritances we leave our descendants: the systems that sustain life in all its forms; the affordances of our cultural and technological heritage; the capacities for care, self-correction and resilience within these.

Above all, the common thread of these delusions is *denial*. They deny our biology and our history; our co-evolution with technology; the plethora of possible futures that await us; the impossibility of innovating our way towards either a perfect society or a perfected version of ourselves. And the antidote to them is, however imperfectly, for us to privilege truth over comfort or dogma: to embrace complexity, compromise and imperfection in a spirit of hope; to strive towards something larger than ourselves without diminishing its payoffs to power or profit.

And Finally

Soon after I became a father, at some point into the first few months of sleepless nights and earlier-than-early mornings, I found myself thinking that my son – my tiny implacable son, who so often I couldn't soothe through his hours of tears – was larger than me. I felt insubstantial, old, starting to break apart; while he had so much time coiled inside him, waiting to become and to keep on becoming.

It was an odd, vertiginous sensation – and also a restatement of the crushingly obvious. There was nothing here I hadn't known for my entire adult life. Birth and death are as basic as it gets. Yet, whenever the universal comes knocking, the fact that it indelibly applies to *us* is still shocking. From the inside, the universe is a resolutely personal and singular experience. One life, one chance to live it, floating in darkness.

In his 2012 book about parenting, children and identity, *Far from the Tree*, the author and psychologist Andrew Solomon writes about the strangeness of this shock – and how a child's stubborn *otherness* can give the lie to the line that we somehow live on through those who come after us:

> Parenthood abruptly catapults us into a permanent relationship with a stranger, and the more alien the stranger, the stronger the

whiff of negativity. We depend on the guarantee in our children's faces that we will not die. Children whose defining quality annihilates that fantasy of immortality are a particular insult.[1]

Even in the tightest embrace of familial love, parents and their children remain divided: inexorably distinct. And the same is true of all forms of inheritance. However desperately we may wish to see in those who come after us a promise of personal continuity, in the end we must love them and let go – or, having failed to pull off the first, nevertheless succumb to the second.

On the outbreak of the Second World War, the poet W. H. Auden – then in his early thirties, recently arrived in New York – conjured in his poem 'September 1, 1939' one of the most resounding dichotomies of twentieth-century poetry: 'we must love one another or die.'[2] It's a line that has commanded desperate assent at times of crisis ever since it was written. It was one of the first responses to the attacks of September 11th read on National Public Radio in America in 2001.[3] Yet its original pairing of love or death, addressed to a world on the brink of war, was one that Auden himself had turned against by the time it came to his 1945 *Collected Poems*. In that edition, he omitted the stanza entirely. By 1964, when Auden was in his late fifties, he altered the line to 'we must love one another and die' when allowing the poem to be reprinted in an anthology. He also insisted that it be accompanied by a note calling it 'trash which [the author] is ashamed to have written.'[4]

Reading Auden when I was a young man, barely out of my teens, I couldn't understand his older self's rejection of his younger self's exaltation. It seemed a strange act of violence, for an elder poet to censor his youth – and perhaps it was. Perhaps he was ashamed to have fled Britain for safe shores. Whatever his reasons, the older Auden was determined to be known for a poetry of honesty; of careful disillusionment rather than dazzled hope.

He was both right and wrong. Right, that we must love one another *and* die: that the two are bound together, and the first is not an alternative to the second. Wrong, that we can reach back through time and rewrite the faults of youth. Even our words escape us, granting immortality not to their author but to whichever ghosts the world wishes their author to have been. As Solomon concludes, perhaps the most difficult task of love is learning to set ourselves aside – and to privilege our descendants' future ahead of our own best hopes:

> We must love them for themselves, and not for the best of ourselves in them, and that is a great deal harder to do. Loving our own children is an exercise for the imagination.[5]

Imagination: that envisioning of a world without us in it, a world we will never see. It is imagination more than philosophy that teaches us how to die: a better task for poets than for philosophers. Or rather, it is imagination that teaches us to think and feel outside the confines of a single life; to see others' lives for what they are; and to love them above whatever we wish them to be.

We must die, this much is certain. Yet nobody and nothing obliges us to love as we do: hopefully, extravagantly, beyond all measure and reason. Love is learned, practiced, handed down; the root of our endurance, the key to our hope. It's not enough, because nothing is. But it is the first and the last thing we have.

Acknowledgements

A few parts of this book have appeared previously, in different forms, across a number of publications. I am deeply grateful both to them and to all those who have commissioned and supported my work over the last decade. The following all appear with kind permission.

Richard Fisher commissioned my February 2019 essay 'Technology in deep time: How it evolves alongside us' for the BBC Future website, elements of which can be found in the first chapter. I am also indebted to Richard's close reading of an early manuscript of this book, not to mention to his friendship, conversation and the richness of his own book, *The Long View* (Wildfire, 2023).

Jemima Kiss commissioned several pieces from me, during her time at *The Guardian*, that helped me discover and deepen my own thinking around technology. In particular, parts of 'Technology is killing the myth of human centrality', from 27 August 2016, can be found in the discussions of 'myth and mystery' and 'coming to terms with our place in the universe' in chapter eight.

Zan Boag, founding editor of *New Philosopher* magazine, has offered me an extraordinary mix of philosophical freedom and opportunity as an author. As a regular contributor to the magazine since 2014, it has been instrumental in helping me write accessibly about themes close to my heart. Aspects of my essay for Issue Six (Progress) were adapted into 'Finding purpose through those who come after us' in chapter twelve; for Issue Seventeen (Communication) into 'Overcoming bullshit' in chapter six; for Issue Twenty-five

(Death) into 'And Finally'; for Issue Thirty (Perception) into the opening of chapter five; for Issue Thirty-Two (Energy) into 'Entering an age of fire' in chapter two; and for Issue Thirty-Five (Love) into parts of the first half of chapter three.

Jonathan Rowson, founding director of the interdisciplinary think-tank Perspectiva, has both inspired me with his own words, work and example, and introduced me to an intellectual and spiritual community of extraordinary richness and generosity. As an Associate at Perspectiva, I first explored several elements of chapters seven and eleven in my February 2021 essay 'Finding Virtue in the Virtual' for Perspectiva. My co-lead for Perspectiva's Digital Ego Project, Dan Nixon, also helped me hone my thinking across a happy host of conversations, exchanges and country walks, as well as deepening my appreciation of phenomenology and mindfulness through his own writing and practices.

Finally, James McConnachie commissioned me to write a piece about authorship and Artificial Intelligence for the Spring 2022 issue of *The Author* magazine, the Journal of the Society of Authors, elements of which are incorporated into 'What do we want from AI?' in chapter nine. I'm grateful to James for the opportunity, as well as for the many happy years we have spent as colleagues working to support authors' rights at ALCS and CLA.

Thanks

Dozens of people have helped this book find its way into the world: the following are just a few of those foremost in my mind. Nothing would or could have happened without the efforts, vision and understanding of my agent, Will Francis; or the encouragement and facilitations of my dear friend Ziyad Marar, who has helped bring about so many of the most marvellous things in my professional life.

Ravi Mirchandani believed in this book enough to take it on, and I'm immensely grateful to him and Mary Mount for helping me bring it to fruition, along with all those at Macmillan and Picador – not least editorial manager Nicholas Blake and jacket designer Tiana Dunlop – who helped transform my thoughts into the finished product before you.

Countless conversations with friends and colleagues have kept me energized and hopeful over the last few years: in particular, Timo Hannay, Rob Poynton, Michael Ridpath, Roman Krznaric and Richard Fisher have given freely of their friendship, attention and insights; while the philosophers David Weinberger, L. M. Sacasas and Evan Selinger have inspired me with their work and examples.

This is, among other things, a book about *reading* – and the enduring power of words on a page to awaken and transport us. Professor John Kelly is one person who opened my eyes to reading in this way, and I owe both him and Christine Kelly a lifelong debt of mentorship and friendship. Behind them both stands my mother's fierce love of books, words and the beauty of language. It is impossible for me to express how much I owe her – or my beloved wife, Cat, and my children, Toby and Clio, who fill my heart and my days with purpose and gratitude.

I offer my deepest thanks to you all – and to *you*, my reader, for coming with me this far. I am conscious that I have ended up writing an immodest book about modesty; a sweeping account of self-aggrandizement's follies. All that I can plead in my defence is that I have tried to find a story worth telling; and that I hope you have found enough truth, beauty or entertainment within it for your time to have been well spent.

Select Bibliography

Anjum R. L. & Mumford S. (2018). *Causation in science and the methods of scientific discovery*. Oxford University Press.

Arthur W. B. (2010). *The nature of technology : what it is and how it evolves*. Penguin.

Baggini J. (2011). *The ego trick*. Granta.

Baggini J. (2020). *The godless gospel : was Jesus a great moral teacher?* Granta.

Beauvoir S. de & O'Brian P. (1968). *Les belles images*. Putnam.

Beauvoir S. de. (1948). *The ethics of ambiguity, translated from the French by Bernard Frechtman*.

Bostrom N. (2014). *Superintelligence : paths, dangers, strategies*. Oxford University Press.

Carr N. G. (2014). *The glass cage : how our computers are changing us*. W. W. Norton & Company.

Castillo E. (2022). *How to read now : essays*. Atlantic Books.

Chalmers D. J. (2022). *Reality plus : virtual worlds and the problems of philosophy*. Penguin Books.

Christian B. (2012). *The most human human : what artificial intelligence teaches us about being alive*. Penguin.

Christian B. (2021). *The alignment problem : how can artificial intelligence learn human values?* Atlantic Books.

Clark A. (2008). *Supersizing the mind : embodiment, action and cognitive extension*. Oxford University Press.

Clark A. (2019). *Surfing uncertainty : prediction action and the embodied mind*. Oxford University Press.

Clarke A. C. (1962). *Profiles of the future : an inquiry into the limits of the possible*. Bantam Books.

SELECT BIBLIOGRAPHY

Crawford K. (2021). *The atlas of AI*. Yale University Press.

Criado Perez C. (2019). *Invisible women : exposing data bias in a world designed for men*. Chatto & Windus.

Dawkins R. (2016). *The selfish gene : 40th anniversary edition*. Oxford University Press.

Dennett D. C. (1991). *Consciousness explained* (1st ed.). Little Brown.

Dennett D. C. (2017). *From bacteria to Bach and back : the evolution of minds*. Allen Lane.

Deutsch D. (2011). *The beginning of infinity : explanations that transform the world*. Penguin.

Didion J. (2005). *The year of magical thinking*. Knopf.

Edmonds D. (2013). *Would you kill the fat man? : the trolley problem and what your answer tells us about right and wrong*. Princeton University Press.

Easwaran E. (1986). *The Bhagavad Gita*. Penguin.

Ellenberg J. (2015). *How not to be wrong : the hidden maths of everyday life*. Penguin Books.

Ellul J. & Wilkinson J. (1964). *The technological society*. Vintage Books.

Finlayson B., Warren G. & Council for British Research in the Levant. (2010). *Landscapes in transition*. Oxbow Books.

Fisher R. (2023). *The long view : why we need to transform how the world sees time*. Wildfire.

Floridi L. (2014). *The 4th revolution : how the infosphere is reshaping human reality*. Oxford University Press.

Floridi L. (2015). *The ethics of information*. Oxford University Press.

Frankfurt H. G. (2005). *On bullshit*. Princeton University Press.

Frischmann B. M. & Selinger E. (2018). *Re-engineering humanity*. Cambridge University Press.

Frith C. D. (2011). *Making up the mind : how the brain creates our mental world*. Blackwell.

Gibson J. J. (1966). *The senses considered as perceptual systems*. Houghton Mifflin.

Gibson J. J. (1979). *The ecological approach to visual perception*. Houghton Mifflin.

Gibson W. & Sterling B. (2011). *The difference engine*. Random House.

SELECT BIBLIOGRAPHY

Gleick J. (2012). *The information : a history, a theory, a flood.* Fourth Estate.

Gopnik A. (2009). *The philosophical baby : what children's minds tell us about truth, love & the meaning of life.* Bodley Head.

Gray J. (2011). *The immortalization commission : science and the strange quest to cheat death.* Allen Lane.

Greene J. D. (2015). *Moral tribes : emotion, reason and the gap between us and them.* Atlantic Books.

Graeber D. & Wengrow D. (2021). *The dawn of everything : a new history of humanity.* Penguin Books

Haidt J. (2012). *The righteous mind : why good people are divided by politics and religion.* Pantheon.

Hall E. (2018). *Aristotle's way : how ancient wisdom can change your life.* Vintage Digital.

Harford T. (2021). *How to make the world add up : ten rules for thinking differently about numbers.* Bridge Street Press.

Held V. (2007). *The ethics of care : personal, political and global.* Oxford University Press.

Hrdy S. B. (2011). *Mothers and others : the evolutionary origins of mutual understanding.* Harvard University Press.

Hoffman D. D. (2020). *The case against reality : how evolution hid the truth from our eyes.* Penguin Books.

Hume D. (1896). *A treatise of human nature.* Clarendon Press.

Illich I. (1978). *The right to useful unemployment and its professional enemies.* Boyars.

Johansson S. & Perry F. (2021). *The dawn of language : how we came to talk.* MacLehose Press.

Kahneman D. (2013). *Thinking, fast and slow.* Farrar, Straus and Giroux.

Kahneman D. , Sibony O. & Sunstein C. R. (2022). *Noise : a flaw in human judgement.* William Collins.

Kant I. & Ellington J. W. (1993). *Grounding for the metaphysics of morals ; with on a supposed right to lie because of philanthropic concerns.* Hackett Publishing.

Kelly K. (2010). *What technology wants.* Viking.

Krznaric R. (2021). *The good ancestor : how to think long-term in a short-term world.* WH Allen.

Kuhn T. S. , Conant J. & Haugeland J. (2000). *The road since structure : philosophical essays 1970-1993 with an autobiographical interview.* University of Chicago Press.

Kuhn T. S. (2009). *The structure of scientific revolutions.* University of Chicago Press.

Kurzweil R. (1999). *The age of spiritual machines : when computers exceed human intelligence.* Viking.

Lanier J. (2019). *Ten arguments for deleting your social media accounts right now.* Vintage.

Le Guin U. K. & Wood S. (1982). *The language of the night : essays on fantasy and science fiction.* Berkley Books.

Le Guin U. K. (1997). *Dancing at the edge of the world : thoughts on words, women, places.* Grove.

Le Guin U. K. (2019). *Words are my matter : writings on life and books.* Mariner Books.

MacAskill W. (2023). *What we owe the future : a million-year view.* Oneworld.

MacIntyre A. C. (2013). *After virtue : a study in moral theory.* Bloomsbury.

MacIntyre A. C. (2009). *Dependent rational animals : why human beings need the virtues.* Open Court.

Marvin C. (1988). *When old technologies were new : thinking about electric communication in the late nineteenth century.* Oxford University Press.

Marcus G. & Davis E. (2019). *Rebooting AI : building artificial intelligence we can trust.* Pantheon Books.

McGilchrist I. (2023). *The matter with things : our brains, our delusions and the unmaking of the world.* Perspectiva Press.

Meijer M. & De Vriese H. (2021). *The philosophy of reenchantment.* Routledge.

Midgley M. (2018). *What is philosophy for?* Bloomsbury Academic.

Morozov E. (2013). *To save everything, click here : technology solutionism and the urge to fix problems that don't exist.* Allen Lane.

Murphy Paul A. (2021) *The Extended Mind : the power of thinking outside the brain.* Houghton Mifflin Harcourt.

Noble S. U. (2018). *Algorithms of oppression : how search engines reinforce racism.* New York University Press.

Nowell A. (2021). *Growing up in the ice age : fossil and archaeological evidence of the lived lives of plio-pleistocene children*. Oxbow Books.

Nye D. E. (2007). *American technological sublime*. MIT Press.

O'Neil C. (2016). *Weapons of math destruction: how big data increases inequality and threatens democracy*. Crown.

Ord T. (2020). *The precipice : existential risk and the future of humanity*. Bloomsbury.

Parfit D. (1984). *Reasons and persons*. Clarendon Press.

Popper K. R. & Eccles J. C. (1984). *The self and its brain*. Routledge & Kegan Paul.

Raihani N. (2022). *The social instinct : how cooperation shaped the world*. Vintage.

Raworth K. (2022). *Doughnut economics : seven ways to think like a 21st-century economist*. Penguin.

Rid T. (2016). *Rise of the machines : the lost history of cybernetics*. Scribe.

Rorty R. (1989). *Contingency, irony and solidarity*. Cambridge University Press.

Rovelli C. (2020). *There are places in the world where rules are less important than kindness*. Allen Lane.

Russell S. (2019). *Human compatible : artificial intelligence and the question of control*. Penguin.

Schatzberg E. (2018). *Technology: critical history of a concept*. University of Chicago Press.

Scheffler S. (2016). *Death and the afterlife*. Oxford University Press.

Seth A. (2021). *Being you : the inside story of your inner universe*. Faber & Faber.

Singer P. (2009). *The life you can save : acting now to end world poverty*. Picador.

Smart J. J. C. & Williams B. (1973). *Utilitarianism : for and against*. Cambridge University Press.

Smith M. R. & Marx L. (1994). *Does technology drive history? the dilemma of technological determinism*. MIT Press.

Solomon A. (2014). *Far from the tree : parents, children and the search for identity*. Vintage.

Standage T. (2003). *The Mechanical Turk : the true story of the chess-playing machine that fooled the world.* Penguin.

Stephenson N. (2012). *Some remarks : essays and other writing.* William Morrow.

Stilgoe J. (2019). *Who's driving innovation? new technologies and the collaborative state.* Palgrave Macmillan.

Taleb N. N. (2007). *Fooled by randomness : the hidden role of chance in life and in the markets.* Penguin.

Taylor T. (2010). *The artificial ape : how technology changed the course of human evolution.* Palgrave Macmillan.

Tetlock P. E. & Gardner D. (2015). *Superforecasting : the art and science of prediction.* Crown.

Thompson E. P. (1963). *The making of the English working class.* Gollancz.

Turkle S. (2017). *Alone together : why we expect more from technology and less from each other.* Basic Books.

Vallor S. (2016). *Technology and the virtues : a philosophical guide to a future worth wanting.* Oxford University Press.

Véliz Carissa. (2020). *Privacy is power : why and how you should take back control of your data.* Bantam Press.

Vernon M. & Dante A. (2021). *Dante's divine comedy : a guide for the spiritual journey.* Angelico Press.

Vince G. (2019). *Transcendence : how humans evolved through fire, language, beauty and time.* Allen Lane.

Virilio P. & Rose J. trans. (2007). *The original accident.* Polity.

Weinberger D. (2019). *Everyday chaos : technology, complexity and how we're thriving in a new world of possibility.* Harvard Business Review Press.

Weizenbaum J. (1976). *Computer power and human reason : steps toward the mechanization of thought.* W.H. Freeman.

Wittgenstein L., Hacker P. M. S. & Schulte J. (2010). *Philosophical investigations* (4th ed.). John Wiley & Sons.

Notes

INTRODUCTION

1 First published in the *News Review* section of *The Sunday Times* on 29 August 1999; available online at https://douglasadams.com/dna/19990901-00-a.html. It's worth noting just how right Adams was about the number of computers. More than a trillion microchips are now made each year, each one a tiny computer. The world would need to boast more than one hundred chairs for every single human alive to equal even this annual total.

2 I owe this framing to a conversation with the author and activist Cory Doctorow, who first discussed it with me during an interview for the *Independent* newspaper, 'Manifesto for a virtual revolution: Cyber-activist Cory Doctorow's new novel imagines a revolt of online slaves', 21 May 2010, online at https://www.independent.co.uk/arts-entertainment/books/features/manifesto-for-a-virtual-revolution-cyberactivist-cory-doctorow-s-new-novel-imagines-a-revolt-of-online-slaves-1978233.html.

3 Culkin was updating a famous phrase coined by Winston Churchill in an October 1943 speech to the House of Commons, referring to the future need to rebuild the debating chamber, which had been destroyed by German bombs. Churchill argued that the chamber must be rebuilt exactly as it had been before, because its layout enshrined Britain's two-party political system: 'We shape our buildings and afterwards our buildings shape us.' See Culkin, J. M. (1967, March), 'A schoolman's guide to Marshall McLuhan', *The Saturday Review*, 51–53, 70–72. Online at https://webspace.royalroads.ca/llefevre/wp-content/uploads/sites/258/2017/08/A-Schoolmans-Guide-to-Marshall-McLuhan-1.pdf. For an account of the debating chamber's destruction and reconstruction see https://www.parliament.uk/about/living-heritage/building/palace/architecture/palacestructure/churchill/.

NOTES

4 The model of 'technological momentum' was first developed by the American historian of technology Thomas P. Hughes, and explores the ways in which technological systems both emerge from particular social contexts and increasingly reshape those contexts over time. As he put it in his essay 'Technological Momentum': 'A technological system can be both a cause and an effect; it can shape or be shaped by society. As they grow larger and more complex, systems tend to be more shaping of society and less shaped by it.' Although, he emphasized, 'technological momentum, like physical momentum, is not irresistible.' Published in *Does Technology Drive History?* eds. Merritt Roe Smith and Leo Marx (MIT Press, 1994), pp. 102–113.

5 This isn't an academic book, but it's worth emphasizing that my arguments are indebted to decades of scholarship in technological history, philosophy and sociology, including – increasingly and importantly – those writing from a diversity of historically under-represented perspectives. I have aimed to acknowledge and quote a selection of relevant sources throughout, and to provide further reading in the bibliography, with an emphasis on work that feels resonant and urgent. Inevitably, I have only gestured towards the richness of these fields, and I would urge you to make my notes and bibliography a starting point for your own enquiries and explorations.

CHAPTER 1: TECHNOLOGY IN DEEP TIME:
THE DELUSION OF INEVITABILITY

1 For a fascinating and comprehensive history of the term 'technology' itself – from its classical origins as a description of 'mere' technical practices to the emergence in the nineteenth century of the German term *Technik* to describe the entire field of material production and its associated expertise – see Eric Schatzberg, *Technology: Critical History of a Concept* (Chicago, 2018).

2 The use of sticks and animal parts is also highly probable, but inherently lacks evidence due to the perishability of these materials. For evidence of the earliest use of stone tools by our ancient ancestors, see Harmand, S., Lewis, J., Feibel, C. et al. (2015), '3.3-million-year-old stone tools from Lomekwi 3, West Turkana, Kenya', *Nature* 521, 310–315. https://doi.org/10.1038/nature14464.

3 See Esteban, Ruth & López, Alfredo & Rios, Álvaro & Ferreira, Marisa & Martinho, Francisco & Méndez Fernandez, Paula & Andréu, Ezequiel

& García, José & Ponzone, Liliana & Ruíz, Rocío & GIL-Vera, Francisco & Bernal, Cristina & Capdevila, Elvira & Sequeira, Marina & Martínez-Cedeira, José. (2022). 'Killer whales of the Strait of Gibraltar, an endangered subpopulation showing a disruptive behavior.' *Marine Mammal Science*. 1–11. 10.1111/mms.1294 https://doi.org/10.1111/mms.12947. Hunt, G. R., and Gray, R. D. (2003), 'Diversification and cumulative evolution in New Caledonian crow tool manufacture', *Proceedings. Biological sciences*, 270(1517), 867–874. https://doi.org/10.1098/rspb.2002.2302.

4 For a fine overview of printing's history before Gutenberg, see 'So, Gutenberg Didn't Actually Invent Printing As We Know It: On the Unsung Chinese and Korean History of Movable Type' by M. Sophia Newman at LitHub.com, 19 June 2019, online at https://lithub.com/so-gutenberg-didnt-actually-invent-the-printing-press.

5 The Morgan Library and Museum in New York City is the only institution in the world to possess three copies of the Gutenberg Bible, and hosts extensive and useful free online resources about the book's history and creation at https://www.themorgan.org/collection/Gutenberg-Bible.

6 Needham, P. (1985), 'The Paper Supply of the Gutenberg Bible', *The Papers of the Bibliographical Society of America*, 79(3), 303–374. http://www.jstor.org/stable/24303663.

7 For a snapshot of just some of the complexities being unearthed by research into agriculture's origins, see Brown, Terence, Jones, Martin K., Powell, Wayne and Allaby, Robin G. (2009), 'The Complex Origins of Domesticated Crops in the Fertile Crescent', *Trends in ecology and evolution*. 24. 103–9. 10.1016/j.tree.2008.09.008.

8 For an in-depth exploration of one of the most famous examples of this phenomenon, the eye, see Lamb, Trevor D., Arendt, Detlev and Collin, Shaun P. (2009), 'The evolution of phototransduction and eyes', *Phil. Trans. R. Soc.* B364: 2791–2793, http://doi.org/10.1098/rstb.2009.0106.

9 For a classic paper evoking some of the unique features of Bronze Age trade, see Ratnagar, S. (2001), 'The Bronze Age: Unique Instance of a PreIndustrial World System?' *Current Anthropology*, 42(3), 351–379, https://doi.org/10.1086/320473.

10 For a wide-ranging and well-illustrated overview of the roots of Egyptian civilization, see Wilkinson, Toby A. H., *Early Dynastic Egypt* (Routledge, 1999).

NOTES

11 For an introduction to the varied evidence of proto-farming activities at Göbekli Tepe, see Curry, Andrew, 'How ancient people fell in love with bread, beer and other carbs', *Nature*, 22 June 2021, online at https://www.nature.com/articles/d41586-021-01681-w – as the author puts it, 'rather than just starting to experiment with wild grains, the monument builders were apparently proto-farmers, already familiar with the cooking possibilities grain offered despite having no domesticated crops.'

12 See his wonderfully detailed and illustrated account of excavations at the site: Schmidt, K. (2010), 'Göbekli Tepe – the Stone Age Sanctuaries. New results of ongoing excavations with a special focus on sculptures and high reliefs', *Documenta Praehistorica*, 37, 239–256. https://doi.org/10.4312/dp.37.21.

13 Terberger, Thomas, Zhilin, Mikhail, Savchenko, Svetlana, 'The Shigir idol in the context of early art in Eurasia', *Quaternary International*, Vol. 573, 2021, pp. 14–29, https://doi.org/10.1016/j.quaint.2020.10.025.

14 Pryor, A., Beresford-Jones, D., Dudin, A., Ikonnikova, E., Hoffecker, J., and Gamble, C. (2020), 'The chronology and function of a new circular mammoth-bone structure at Kostenki 11', *Antiquity*, 94(374), 323–341. doi:10.15184/aqy.2020.7.

15 For some of the foundational research in this area, see T. Watkins, 'Changing People, Changing Environments: How Hunter-Gatherers Became Communities that Changed the World' in B. Finlayson, G. Warren (eds.), *Landscapes in Transition* (Oxbow Books, 2010), pp. 106–114. For an influential recent account of such sites, and speculation as to their significance, see David Graeber and David Wengrow, *The Dawn of Everything: A New History of Humanity* (Penguin, 2021), and in particular pp. 89–91. It's also worth noting the remarkable recent unearthing near the border between Zambia and Tanzania of what is thought to be the oldest wooden structure in the world. Dated to almost half a million years ago, it's a discovery that suggests remarkable cultural and technological sophistication among proto-humans at that time. See Barham, L., Duller, G. A. T., Candy, I. et al. 'Evidence for the earliest structural use of wood at least 476,000 years ago.' *Nature* (2023). https://doi.org/10.1038/s41586-023-06557-9.

16 See Zeder, Melinda A., 'The Origins of Agriculture in the Near East', *Current Anthropology* 2011 52:S4, S221–S235, https://doi.org/10.1086/659307.

17 The line comes from the preface to E. P. Thompson's 1963 *The Making of the English Working Class*, in which he described his intention 'to rescue

NOTES

the poor stockinger, the "obsolete" hand-loom weaver, the "Utopian" artisan, and even the deluded follower of Joanna Southcott, from the enormous condescension of posterity'.

18 For a work that helped to establish the steampunk genre, see William Gibson and Bruce Sterling's 1990 novel *The Difference Engine* (Gollancz), which imagines a parallel Victorian era in which Charles Babbage successfully built his mechanical programmable computer, the titular difference engine, and the information revolution thus played out in an era of steam and mechanical computation.

19 G. E. Moore, 'Cramming more components onto integrated circuits, Reprinted from *Electronics*, volume 38, number 8, April 19, 1965, pp.114 ff.,' in IEEE Solid-State Circuits Society Newsletter, vol. 11, no. 3, pp. 33–35, September 2006, doi: 10.1109/N-SSC.2006.4785860.

20 On release in 1979, the Motorola 68000 had the highest transistor count of any mainstream commercial chip: around 70,000.

21 See https://en.wikipedia.org/wiki/Transistor_count for a regularly updated record of past and present transistor counts.

22 Leiserson, C. E., Thompson, N. C., Emer, J. S., Kuszmaul, B. C., Lampson, B. W., Sanchez, D., and Schardl, T. B. (2020). 'There's plenty of room at the Top: What will drive computer performance after Moore's law?'. *Science* (New York, N.Y.), 368(6495), eaam9744. https://doi.org/10.1126/science.aam9744

23 Ray Kurzweil, 'The Law of Accelerating Returns', 7 March 2001, online at https://www.kurzweilai.net/the-law-of-accelerating-returns.

24 Ibid.

CHAPTER 2: STONE AND FIRE: THE DELUSION OF MASTERY

1 See 'On the Origins of "The March of Progress"' by Kevin Blake, 17 December 2018, online at https://sites.wustl.edu/prosper/on-the-origins-of-the-march-of-progress/.

2 *Hominins* is a subtly different term to the more familiar term *hominids*. *Hominins* encompasses both modern humans and all of our extinct immediate kin; while *hominids* is a larger category encompassing all living and extinct great apes, including ourselves, chimpanzees, bonobos, orangutans and gorillas.

3 There's plenty of controversy and uncertainty around hominins' ancient origins, not least around the question of whether these lie entirely in Africa. Hominin-like footprints found in Crete, for example, may place

very early hominins in Eurasia some six million years ago. What does seem clear, however, is that our more immediate ancestors – and *Homo sapiens* itself – evolved within and ranged across the African continent. For some recent evidence and an account of the debate, see Kirscher, U., El Atfy, H., Gärtner, A. et al. (2021), 'Age constraints for the Trachilos footprints from Crete', *Sci Rep 11*, 19427, https://doi.org/10.1038/s41598-021-98618-0.

4 McPherron, S., Alemseged, Z., Marean, C. et al. (2010), 'Evidence for stone-tool-assisted consumption of animal tissues before 3.39 million years ago at Dikika, Ethiopia', *Nature* 466, 857–860, https://doi.org/10.1038/nature09248.

5 See Ward, C. V. and Hammond, A. S. (2016), 'Australopithecus and Kin', *Nature Education Knowledge* 7(3):1.

6 Antón, S. C., 'All who wander are not lost', *Science* 2020; 368: 34–35, doi: 10.1126/science.abb4590.

7 Whether *Homo habilis* and *Homo erectus* were the only two ancient members of *Homo* is much debated and disputed, largely because the paleontological evidence is so scanty – and the state of our knowledge continues to be updated in the light of new discoveries and theories.

8 See Torralvo, Kelly, Rabelo, Rafael, Andrade, Alfredo and Botero-Arias, Robinson (2017), 'Tool use by Amazonian capuchin monkeys during predation on caiman nests in a high-productivity forest', *Primates*. 58: 279–283. 10.1007/s10329-017-0603-1.

9 See Julian K. Finn, Tom Tregenza and Mark D. Norman (2009), 'Defensive tool use in a coconut-carrying octopus', *Current Biology*, Volume 19, Issue 23, https://doi.org/10.1016/j.cub.2009.10.052. In terms of tool usage by the other animals mentioned in this paragraph, elephants put logs and branches to a wide range of uses, including knocking down obstacles and swatting flies; sea otters use rocks for breaking open shells; while dolphins have been observed carrying marine sponges to help them flush out prey on the ocean floor. And don't forget about the Caledonian crows.

10 See de la Torre, I. (2016), 'The origins of the Acheulean: past and present perspectives on a major transition in human evolution', *Philosophical Transactions of the Royal Society of London. Series B, Biological sciences*, 371(1698), 20150245, https://doi.org/10.1098/rstb.2015.0245.

11 Arthur, W. Brian, *The Nature of Technology* (Penguin, 2010), p. 23.

12 For example see Gallotti, R., Muttoni, G., Lefèvre, D. et al. (2021), 'First high resolution chronostratigraphy for the early North African Acheulean

at Casablanca (Morocco)', *Sci Rep* 11, 15340, https://doi.org/10.1038/s41598-021-94695-3.

13 Mussi, M., Mendez-Quintas, E., Barboni, D. et al. (2023), 'A surge in obsidian exploitation more than 1.2 million years ago at Simbiro III (Melka Kunture, Upper Awash, Ethiopia)'. *Nat Ecol Evol* 7, 337–346, https://doi.org/10.1038/s41559-022-01970-1.

14 For instance, there are those who have argued that Acheulean industry may have been driven by innate rather than cultural impulses, based in part upon its lack of the kind of variations associated with cultural evolution. See Corbey, R., Jagich, A., Vaesen, K., and Collard, M. (2016), 'The Acheulean handaxe: More like a bird's song than a Beatles' tune?'. *Evolutionary anthropology*, 25(1), 6–19, https://doi.org/10.1002/evan.21467. I'm not convinced by this line of argument – but the potential strength of genetic influence on early toolmaking is well worth noting. For another interesting discussion of the relationship between the 'imposed' and 'emergent' features of hand-axe manufacture, and how these need not represent an either/or dichotomy, see Hutchence, L. and Scott, C. (2021), 'Is Acheulean Handaxe Shape the Result of Imposed "Mental Templates" or Emergent in Manufacture? Dissolving the Dichotomy through Exploring "Communities of Practice" at Boxgrove, UK', *Cambridge Archaeological Journal*, 31(4), 675–686, doi:10.1017/S0959774321000251.

15 For a summary of some of these debates see Spikins, Penny (2012), 'Goodwill Hunting?: Debates over the "meaning" of Lower Palaeolithic handaxe form revisited', *World Archaeology*. 44:3, 378–392. ISSN 1470-1375, https://doi.org/10.1080/00438243.2012.725889.

16 Suddendorf, T., Addis, D. R., Corballis, M. C., 'Mental time travel and the shaping of the human mind', *Phil Trans R Soc Lond B Biol Sci*. 2009 May 12; 364(1521):1317–1324, doi: 10.1098/rstb.2008.0301. PMID: 19528013; PMCID: PMC2666704. Online at https://www.ncbi.nlm.nih.gov/pmc/articles/PMC2666704/.

17 For a rich account of the varied recent research into the global spread of *Homo erectus* and its remarkable global success and endurance, see Josie Glausiusz, 'What Drove Homo Erectus Out of Africa?', *Smithsonian Magazine*, 19 October 2021, online at https://www.smithsonianmag.com/science-nature/what-drove-homo-erectus-out-of-africa-180978881/.

18 King James Bible, Genesis 1:26. It's worth noting that, in the larger scheme of things, the monotheistic narrative of Genesis is extremely 'modern' and was probably preceded by many millennia of animistic and polytheistic practices that placed humanity far less at the centre of things.

19 Recent research continues to complicate the picture of human evolution, particularly when it comes to the status and usefulness of *heidelbergensis* as a hominin taxon: it has been suggested this term should be suppressed and replaced by the more precise term *Homo bodoensis* to describe our immediate ancestors from between around 900,000 and 500,000 years ago. See Roksandic, M., Radović, P., Wu, X.-J., Bae, C. J., 'Resolving the "muddle in the middle": The case for *Homo bodoensis sp. nov.*', *Evolutionary Anthropology*. 2021; 1–10. https://doi.org/10.1002/evan.21929.

20 Huff, C. D., Xing, J., Rogers, A. R., Witherspoon, D., and Jorde, L. B. (2010), 'Mobile elements reveal small population size in the ancient ancestors of Homo sapiens'. *Proceedings of the National Academy of Sciences of the United States of America*, 107(5), 2147–2152. https://doi.org/10.1073/pnas.0909000107. The authors argue that 'by comparing the likelihood of various demographic models, we estimate that the effective population size of human ancestors living before 1.2 million years ago was 18,500, and we can reject all models where the ancient effective population size was larger than 26,000'.

21 Hublin, J. J., Ben-Ncer, A., Bailey, S. et al. (2017), 'New fossils from Jebel Irhoud, Morocco and the pan-African origin of *Homo sapiens*', *Nature* 546, 289–292, https://doi.org/10.1038/nature22336.

22 For a recent perspective on the waves of 'out of Africa' hominin migrations, integrating genetic with paleontological evidence, see Leonardo Vallini, Giulia Marciani, Serena Aneli, Eugenio Bortolini, Stefano Benazzi, Telmo Pievani, Luca Pagani, 'Genetics and Material Culture Support Repeated Expansions into Paleolithic Eurasia from a Population Hub Out of Africa', *Genome Biology and Evolution*, Volume 14, Issue 4, April 2022, evac045, https://doi.org/10.1093/gbe/evac045.

23 Osipov, S., Stenchikov, G., Tsigaridis, K. et al. (2021), 'The Toba supervolcano eruption caused severe tropical stratospheric ozone depletion'. *Commun Earth Environ* 2, 71, https://doi.org/10.1038/s43247-021-00141-7.

24 Wolf, A. B. and Akey, J. M. (2018). 'Outstanding questions in the study of archaic hominin admixture.' *PLoS genetics*, 14(5), e1007349. https://doi.org/10.1371/journal.pgen.1007349.

25 Sverker Johansson, trans. Frank Perry, *The Dawn of Language: The story of how we came to talk* (Hachette, 2021).

26 See P. Raia et al. (2020), 'Past extinctions of *Homo* species coincided with increased vulnerability to climatic change', *One Earth*, 3:4, 480–490.

https://doi.org/10.1016/j.oneear.2020.09.007. Online at https://www.cell.com/one-earth/fulltext/S2590-3322(20)30476-0.

27 Gaia Vince, *Transcendence: How Humans Evolved through Fire, Language, Beauty, and Time* (Allen Lane, 2019), p. 20.

28 Mark Bonta, Robert Gosford, Dick Eussen, Nathan Ferguson, Erana Loveless, and Maxwell Witwer, 'Intentional Fire-Spreading by "Firehawk" Raptors in Northern Australia', *Journal of Ethnobiology* 37(4), 700–718 (1 December 2017), https://doi.org/10.2993/0278-0771-37.4.700.

29 For further details of the evidence discussed in this paragraph, and a brilliant narrative of our ancestral relationship with fire, see Gowlett, J. A. J. (2016), 'The discovery of fire by humans: a long and convoluted process', *Phil. Trans. R. Soc. B: Biological Sciences*. 371 (1696): 20150164, doi:10.1098/rstb.2015.0164.

30 Gaia Vince, *Transcendence*, p. 22.

31 Ibid., p. 9.

32 In one of the earliest versions of the Prometheus myth, told by the Greek poet Hesiod around 700 BCE, the Titan Prometheus steals fire from Zeus through trickery in order to give it to humanity. Hesiod offers two tales of the resulting punishment. In one, Prometheus is chained to a mountain for all eternity, an eagle visiting him each day to eat the ever-regenerating liver from his immortal body. In the other, Zeus creates a woman called Pandora and sends her to Earth where, in her curiosity, she opens the lid of a great jar that turns out to contain the evils of disease, hard labour and suffering. Superhuman power, in other words, comes at a terrible price both for those who take and who wield it.

33 As measured by the US National Oceanic and Atmospheric Administration at the Mauna Loa Observatory in Hawaii. Pre-industrial levels were around 280 parts per million; in general, atmospheric carbon dioxide has fluctuated with glacial cycles over the last million years between approximately 175 and 275 parts per million. See https://climate.nasa.gov/vital-signs/carbon-dioxide/.

34 The term 'pyrocene' was coined by Pyne in a May 2015 essay for *Aeon* magazine, 'The Fire Age', online at https://aeon.co/essays/how-humans-made-fire-and-fire-made-us-human. This passage is taken from his follow-up essay from November 2019, 'The planet is burning', online at https://aeon.co/essays/the-planet-is-burning-around-us-is-it-time-to-declare-the-pyrocene.

35 Pyne, 'The planet is burning'.

NOTES

CHAPTER 3: LOVE AND LEARNING: THE DELUSION OF BRUTALITY

1 For some fascinating comparisons between human and primate reproduction in the evolutionary context, see Martin, R. D. (2007), 'The evolution of human reproduction: a primatological perspective', *American Journal of Biological Anthropology*, Suppl 45, 59–84, https://doi.org/10.1002/ajpa.20734.

2 Alison Gopnik, *The Philosophical Baby: What Children's Minds Tell Us about Truth, Love and the Meaning of Life* (Bodley Head, 2009), p. 8.

3 For a thorough review of much of the necessarily uncertain evidence around the emergence of infant carrying and associated technologies, see Berecz, B., Cyrille, M., Casselbrant, U., Oleksak, S., Norholt, H., 'Carrying human infants – An evolutionary heritage', *Infant Behavior and Development*. August 2020, 60. 101460, https://doi: 10.1016/j.infbeh.2020.101460.

4 Timothy Taylor, *The Artificial Ape: How Technology Changed the Course of Human Evolution* (Palgrave Macmillan, 2010), p. 8.

5 Sarah Blaffer Hrdy, *Mothers and Others: The Evolutionary Origins of Mutual Understanding* (Harvard, 2009), pp. 3–4.

6 Theories as to how and why the menopause evolved in humans vary considerably, ranging from the benefits of support from post-reproductive females in terms of kin selection to its effectively accidental emergence as a product of recent extensions in human lifespans. One theory combining several of these elements can be found in Takahashi, M., Singh, R. S. and Stone J. (2017), 'A Theory for the Origin of Human Menopause', *Front. Genet.* 7:222, doi: 10.3389/fgene.2016.00222, which suggests that 'mating involving only young women allow[ed] late-onset fertility-diminishing alleles to accumulate cryptically. Increased lifespan necessarily allowed reproductive senescence to be expressed. Grandmothers may have emerged as a means to reclaim lost fitness through kin selection. Menopause thus theoretically may be considered a trait that originated neutrally, evolved adaptively, and was co-opted exaptively.' While debate is sure to continue, what we can say for certain is that the extent of the human menopause is both unique among primates and vanishingly rare among mammals. As the above paper puts it: 'if menopause is considered quantitatively on the basis of life history as a proportionately long time in adulthood during which individuals are non-reproductive . . . then it indeed is remarkable in humans relative to

other species in the animal kingdom, occurring also only in killer whales . . . and perhaps short-finned pilot whales.' Orcas, interestingly, are noteworthy for the fact that the presence of post-reproductive females seems substantially to enhance their grand-offspring's survival: see Nattrass, S., et al. (20190, 'Postreproductive killer whale grandmothers improve the survival of their grandoffspring', *Proceedings of the National Academy of Sciences of the United States of America* 116:26669–26673, https://doi.org/10.1073/pnas.1903844116.
7 Hrdy, *Mothers and Others*, pp. 20–1.
8 In fact, anatomizing different kinds of love was once considered a central task for philosophy. The Ancient Greeks had (at least) seven distinct words for 'love', spanning the public and private domains as well as a gamut of intensities and responsibilities: *agápē* (selfless or unconditional love, such as for one's children, humanity as a whole, or a deity), *érōs* (intimate or erotic passion), *philía* (loyal and affectionate friendship), *philautía* (regard for oneself, or for the intrinsic pleasures of undertaking a task), *storgē* (customary or appropriate affection towards institutions such as the family or state) and *xenía* (hospitality and generosity towards guests or strangers).
9 There are plenty of recent accounts that *do* engage with the primacy of sentiment and cooperation, of course. As the American philosopher and neuroscientist Joshua Greene, for example, writes in his 2013 book *Moral Tribes*, 'familial love is more than just a warm and fuzzy thing. It's a strategic biological device, a piece of moral machinery that enables genetically related individuals to reap the benefits of cooperation.'
10 Gopnik, *The Philosophical Baby*, p. 11.
11 For a statistical analysis of the degree to which women are effaced in countless contemporary contexts, and the tech industry in particular, see Caroline Criado Perez's remarkable book *Invisible Women: Exposing Data Bias in a World Designed for Men* (Chatto and Windus, 2019). As she put it in a July 2019 interview with *Wired* magazine, 'I always think of Sheryl Sandberg going in to ask the head of Google to put in pregnancy parking [spaces] and he said, *I never thought about it* . . .' online at https://www.wired.com/story/caroline-criado-perez-invisible-women/.
12 For the most famous example of Darwin's deep interest in collaboration, generosity and selflessness as group survival traits, see *The Descent of Man* (1871), Chapter Five, which discusses the 'moral faculties' in the context of evolution.
13 Alison Gopnik, *The Philosophical Baby*, p. 10.

14 For a useful outline of the history of humans in Britain, see 'Migration Event: When did the first humans arrive in Britain?' by Josie Mills for *UCL Researchers in Museums*, 24 February 2019, online at https://blogs.ucl.ac.uk/researchers-in-museums/2019/02/24/migration-event-when-did-the-first-humans-arrive-in-britain/.

15 Eiberg, H., Troelsen, J., Nielsen, M. et al. (2008), 'Blue eye color in humans may be caused by a perfectly associated founder mutation in a regulatory element located within the *HERC2* gene inhibiting *OCA2* expression', *Hum Genet* 123, 177–187, https://doi.org/10.1007/s00439-007-0460-x.

16 J. Burger, M. Kirchner, B. Bramanti, W. Haak, M. G. Thomas, 'Absence of the lactase-persistence-associated allele in early Neolithic Europeans.' *Proceedings of the National Academy of Sciences* Mar 2007, 104 (10) 3736–3741; doi: 10.1073/pnas.0607187104.

17 Leda Cosmides and John Tooby, *Evolutionary Psychology: A Primer* (1997), archived at https://web.archive.org/web/20230206213533/https://www.cep.ucsb.edu/primer.html (the original is no longer online).

18 April Nowell, 'Children of the Ice Age', *Aeon*, 13 February 2023, online at https://aeon.co/essays/what-was-it-like-to-grow-up-in-the-last-ice-age; and for a more in-depth exploration, see Nowell's book *Growing Up in the Ice Age* (Oxbow, 2021). As Nowell notes, the archaeology of childhood owes a great deal to the pioneering work of Norwegian archaeologist Grete Lillehammer; to later researchers such as Traci Ardren, Kathryn Kamp and Jane Eva Baxter; and to many others, credited in her book, who continue to build upon their work.

19 Nowell, 'Children of the Ice Age'.

20 I have taken this vivid account from a Cambridge University news article describing part of the 2011 Archaeology of Childhood conference there; see 'Archaeological research reveals that 13,000 years before CBeebies hunter-gatherer children as young as three were creating art in deep, dark caves alongside their parents' 30 September 2011, online at https://www.cam.ac.uk/research/news/prehistoric-pre-school. For the foundational work in this area, see Sharpe, K., and Van Gelder, L. (2006), 'The Study of Finger Flutings', *Cambridge Archaeological Journal*, 16(3), 281–295, doi:10.1017/S0959774306000175.

21 Nichola Raihani, *The Social Instinct: How Cooperation Shaped the World* (Vintage, 2022), p. 2.

22 Raihani offers a fascinating analysis of some of the most significant evidence around primate sensitivity to inequity in footnote 8 to Chapter 14 of *The Social Instinct* on pp. 279–280.
23 McAuliffe, K., Blake, P. R., and Warneken, F. (2014), 'Children reject inequity out of spite', *Biology letters*, 10(12), 20140743, https://doi.org/10.1098/rsbl.2014.0743.
24 A huge amount of research has explored this area over the last few decades, with key papers including: Fehr, E. and Schmidt, K. M. (1999), 'A theory of fairness, competition, and cooperation', *Q. J. Econ.* 114, 817–868, doi:10.1162/003355399556151; Loewenstein, G. F., Thompson, L. and Bazerman, M. H. (1989), 'Social utility and decision making in interpersonal contexts', *J. Pers. Soc. Psychol.* 57(3), 426–441, doi:10.1037/0022-3514.57.3.426; Dawes, C. T., Fowler, J. H., Johnson, T., McElreath, R. and Smirnov, O. (2007). 'Egalitarian motives in humans.' *Nature* 446, 794–796, doi:10.1038/nature05651; Güth, W., Schmittberger, R. and Schwarze, B. (1982), 'An experimental analysis of ultimatum bargaining', *J. Econ. Behav. Organ.* 3(4), 367–388, doi:10.1016/0167-2681(82)90011-7.
25 For a sophisticated analysis of research into fairness via the so-called Ultimatum Game – and the complexities associated with the existence of both 'prosocial' and 'spiteful' punishers – see Brañas-Garza, P., Espín, A., Exadaktylos, F. et al. (2014), 'Fair and unfair punishers coexist in the Ultimatum Game', *Sci Rep* 4, 6025, https://doi.org/10.1038/srep06025.
26 Richard Rorty, *Contingency, Irony, and Solidarity* (CUP, 1989), p. 6.

CHAPTER 4: EXTENDED MINDS: THE DELUSION OF 'IT' AND 'US'

1 Andy Clark, David Chalmers, 'The Extended Mind', *Analysis*, Volume 58, Issue 1, January 1998, pp. 7–19, https://doi.org/10.1093/analys/58.1.7.
2 If using a technology makes no difference to your mind and experiences, why would this choice matter?
3 See David Chalmers, *Reality+* (Penguin, 2022), p. 296.
4 Ibid., p. 297. It's worth adding here one of the most famous formulations of such a sentiment, attributed by the philosopher Karl Popper in his 1966 essay 'On Clouds and Clocks' (first presented as a lecture in 1965) to the physicist Albert Einstein: 'we use, and build, computers because they can do many things which we cannot do; just as I use a pen or pencil when I wish to tot up a sum I cannot do in my head. "My pencil is more intelligent than I", Einstein used to say.' There's no direct evidence that Einstein actually said this, but Popper's crediting of the phrase to him has

stuck – and it is a fine summary of the central claim of the extended mind hypothesis.

5 NB: one alternative view to the extended mind hypothesis is that of *embedded cognition*, which puts a more modest claim at its heart: that, rather than being literally extended, human minds have their capacities enlarged by their embedding in particular environments. According to this view, the mind itself remains inside the body; while the focus of analysis shifts to a constant negotiation between internal cognitive processes and external artefacts, and between the body and its environment. As Chalmers notes in his introduction to Clark's 2008 book *Supersizing the Mind*, this is to some extent a distinction without a difference – or, at least, an opportunity to switch between frames of reference depending on your focus and purposes: 'We have a sort of Necker Cube effect, with mental states counting as extended or not depending on our perspective and our purposes.' Introduction to Andy Clark, *Supersizing the Mind: Embodiment, Action and Cognitive Extension* (OUP, 2008), pp. xii–xiii.

6 Brem, S., Bach, S., Kucian, K., Kujala, K. V., Guttorm, T. K., Martin, E., Lyytinen, H., Brandeis, D., Richardson, U., 'Brain sensitivity to print emerges when children learn letter–speech sound correspondences', *Proc Natl Acad Sci USA*. 2010 Apr 15; 107(17):7939–7944, doi: 10.1073/pnas.0904402107.

7 For an exploration of just some of the ways reading and writing transform the mind, see Falk Huettig, Régine Kolinsky and Thomas Lachmann (2018), 'The culturally co-opted brain: how literacy affects the human mind', *Language, Cognition and Neuroscience*, 33:3, 275–277, doi: 10.1080/23273798.2018.1425803, and the detailed articles to which this piece serves as a preface.

8 For a well-illustrated history of the earliest alphabets, see 'Alphabet Origins: From Kipling to Sinai', 22 December 2014, by Elizabeth Knott for the Met Museum website at https://www.metmuseum.org/exhibitions/listings/2014/assyria-to-iberia/blog/posts/alphabet.

9 According to OECD and UNESCO data aggregated at https://ourworldindata.org/literacy a majority of the world's adult population were illiterate as recently as 1940 (58% illiterate versus 42% literate).

10 This and subsequent quotes are taken from Benjamin Jowett's classic translation, which can be read for free online in full at http://classics.mit.edu/Plato/phaedrus.html. I haven't followed the practice of putting the definite article 'the' in front of each dialogue's name because, to those

NOTES

unfamiliar with it, I feel it can make mentions of them feel unnecessarily stilted.

11 Ibid.
12 James Gleick, *The Information: A History, A Theory, A Flood* (Fourth Estate, 2012), pp. 49–50.
13 Once again, I have used Jowett's translation, which can be read in full online for free at http://classics.mit.edu/Plato/gorgias.1b.txt.
14 Ibid.
15 Plato, *The Republic*, Book X, once again in Benjamin Jowett's translation, online in full at http://classics.mit.edu/Plato/republic.11.x.html. I have used the definite article for the title of *The Republic* because, in this case, I feel it would read more awkwardly to leave it out.
16 For an interesting discussion of Plato as a philosopher of information technology, see Luciano Floridi in conversation with Nigel Warburton for the FiveBooks website, online at https://fivebooks.com/best-books/luciano-floridi-philosophy-information/. For a full text of Plato's *Republic* online see http://classics.mit.edu/Plato/republic.html (the parable of the cave is in Book VII).
17 On the other hand, as anyone who has watched a recent thriller will know, discarding your phone is essential if you don't want the baddies to know where you are.
18 Andy Clark, David Chalmers, 'The Extended Mind'.
19 Alison Gopnik, *The Philosophical Baby*, p. 3.
20 Daniel Kahneman, *Thinking, Fast and Slow* (Farrar, Straus and Giroux, 2011), pp. 50–51.
21 For example, see Janssen, I., LeBlanc, A. G. (2010), 'Systematic review of the health benefits of physical activity and fitness in school-aged children and youth', *Int J Behav Nutr Phys Act* 7, 40, https://doi.org/10.1186/1479-5868-7-40.
22 Annie Murphy Paul, *The Extended Mind: The Power of Thinking Outside the Brain* (Houghton Mifflin Harcourt, 2021), pp. x and 15–16.
23 Andy Clark, David Chalmers, 'The Extended Mind'.
24 UN Human Rights Council, Thirty-second session, Agenda item 3, 'Promotion and protection of all human rights, civil, political, economic, social and cultural rights, including the right to development.' 27 June 2016, United Nations General Assembly document no. A/HRC/32/L.20 ORAL REVISIONS 30 June, online at https://www.article19.org/data/files/Internet_Statement_Adopted.pdf.
25 Annie Murphy Paul, *The Extended Mind*, p. 17.

NOTES

CHAPTER 5: CONSCIOUSNESS AS CONTROLLED HALLUCINATION: THE DELUSION OF LITERAL-MINDEDNESS

1 Müller-Lyer, F. C. (1889), 'Optische Urteilstäuschungen', *Archiv für Physiologie Suppl.*: 263–270. Online at https://www.biodiversitylibrary.org/page/35372625#page/275/mode/1up.
2 Pressey, A. W., Di Lollo, V., and Tait, R. W. (1977), 'Effects of gap size between shaft and fins and of angle of fins on the Müller-Lyer illusion'. *Perception*, 6(4), 435–439, https://doi.org/10.1068/p060435.
3 Anil Seth, *Being You* (Faber, 2021), digital edition, location 1,389.
4 Donald Hoffman, *The Case Against Reality: How Evolution Hid the Truth from Our Eyes* (Penguin, 2019), pp. xii–xiii.
5 A term coined by the philosopher Daniel Dennett as a critique of the seventeenth-century philosopher René Descartes's suggestion that our intangible souls play the part of an inner observing 'us'. See Dennett's 1991 book *Consciousness Explained* (Little, Brown), p. 107: 'Cartesian materialism is the view that there is a crucial finish line or boundary somewhere in the brain, marking a place where the order of arrival equals the order of "presentation" in experience because what happens there is what you are conscious of . . . Many theorists would insist that they have explicitly rejected such an obviously bad idea. But . . . the persuasive imagery of the Cartesian Theater keeps coming back to haunt us – laypeople and scientists alike – even after its ghostly dualism has been denounced and exorcized.'
6 Although it doesn't contain the precise phrase 'controlled hallucination', Frith's book *Making Up the Mind* (Blackwell, 2007) offers a detailed exposition of how, as the title of its fifth chapter puts it, 'Our perception of the world is a fantasy that coincides with reality'. Andy Clark is among the others to have popularized the concept of controlled hallucination in, for example, *Surfing Uncertainty: Prediction, Action, and the Embodied Mind* (OUP, 2015).
7 Anil Seth, *Being You*, digital edition, location 1,425.
8 Ibid., locations 1,443 and 1,475.
9 For a classic example of just some of the complexities bound up with this concept, see Anderson, J. and Fisher, B. (1978), 'The Myth of Persistence of Vision', *Journal of the University Film Association*, 30(4), 3–8, http://www.jstor.org/stable/20687445, and the follow-up article, Anderson, J. and Anderson, B. (1993), 'The Myth of Persistence of Vision Revisited',

Journal of Film and Video, 45(1), 3–12. http://www.jstor.org/stable/20687993.
10 David Deutsch, *The Beginning of Infinity* (Penguin, 2011), p. 38.
11 Ibid., p. 40.
12 David Hume, *A Treatise of Human Nature* (Oxford, 1896), Book 1, Part 4, Section 6, online at https://oll.libertyfund.org/title/bigge-a-treatise-of-human-nature#Hume_0213_642.
13 Julian Baggini, *The Ego Trick* (Granta, 2011), p. 121.
14 For an influential discussion of this and related problems, see Wittgenstein's *Philosophical Investigations*, Part 2, Section XI. As he puts it, 'We find certain things about seeing puzzling, because we do not find the whole business of seeing puzzling enough.'
15 Iain McGilchrist, *The Matter With Things: Our Brains, Our Delusions, and the Unmaking of the World* (Perspectiva Press, 2021), vol. 1, p. 8.
16 There's a fine discussion of the *Upanishads* in the context of Hume and physics in the introduction to the Penguin Classics edition of *The Bhagavad Gita* by Eknath Easwaran. See *The Bhagavad Gita* (Penguin, 1996), translated with a general introduction by Eknath Easwaran, with chapter introductions by Diana Morrison.
17 See Charlie Warzel, 'I Talked to the Cassandra of the Internet Age', *New York Times*, 4 February 2021, online at https://www.nytimes.com/2021/02/04/opinion/michael-goldhaber-internet.html.
18 I'm indebted to the author Robert Poynton for helping me to crystallize these thoughts, and for his gift of these two phrases: 'Attention is where we live. What I pay attention to becomes my life.'
19 McGilchrist, I. (2011), 'Paying attention to the bipartite brain', *Lancet* (London, England), 377(9771), 1068–1069, https://doi.org/10.1016/s0140-6736(11)60422-4.

CHAPTER 6: HOW TECHNOLOGIES INVENT THEMSELVES: THE DELUSION OF COMPREHENSION

1 Daniel Dennett, *From Bacteria to Bach and Back* (Allen Lane, 2017), pp. 147–8.
2 Ibid., p. 213. Dennett is drawing on a 2008 study of the evolution of Polynesian canoes – Deborah S. Rogers, Paul R. Ehrlich, 'Natural selection and cultural rates of change', *Proceedings of the National Academy of Sciences*, March 2008, 105 (9) 3416–3420, doi: 10.1073/pnas.0711802105a – which contains this translation of a passage from a 1908

paper by Émile-Auguste Chartier (most often known by his pen-name Alain) taken from *Propos d'un Normand 1906–1914* (Gallimard, 1956).
3 See in particular his in-depth treatment of this theme in *From Bacteria to Bach*, pp. 56–9.
4 Consider, respectively, the invention of lasers in 1958 and the ongoing achievements of modern Artificial Intelligence.
5 Daniel Dennett, *From Bacteria to Bach and Back*, pp. 312–13.
6 Jordan Ellenberg, *How Not To Be Wrong* (Penguin, 2015), digital edition, location 143.
7 Nassim Nicholas Taleb, *Fooled By Randomness* (Penguin, 2007), digital edition, locations 854 to 893.
8 Tim Harford, *How to Make the World Add Up* (Bridge Street Press, 2020), p. 87.
9 See, for example, this admiring analysis of Wald's achievement from the perspective of the 1980s, and in particular its praise of the sophistication of his insights in the absence of almost any pre-existing mathematical techniques for modelling risks of the kind under consideration: Mangel, Marc; Samaniego, Francisco J. (June 1984), 'Abraham Wald's Work on Aircraft Survivability', *Journal of the American Statistical Association* 79 (386): 259–267, doi:10.1080/01621459.1984.10478038. Online at https://people.ucsc.edu/~msmangel/Wald.pdf.
10 Abraham, Wald (1980) [1943], A Reprint of 'A Method of Estimating Plane Vulnerability Based on Damage of Survivors' (PDF) (Technical report). Center for Naval Analyses; Statistical Research Group, National Defense Research Committee. ADA091073 – via Defense Technical Information Center. Online at https://apps.dtic.mil/sti/pdfs/ADA091073.pdf.
11 To quote some personal correspondence with the author Robert Poynton on this point: 'Improvement is as much about *noticing* as imagining. Notice what works, do it more or again. It doesn't require you to have intended anything. Just paid attention.'
12 Paul Virilio trans. Julie Rose, *The Original Accident* (Polity Press, 2007), pp. 10–11, first published in French as *L'accident original* (Éditions Galilée, 2005).
13 This phrase served as Facebook's internal motto until 2014, and was indebted to the theory of 'disruptive innovation' coined by authors Clayton Christensen and J. L. Bower in their 1995 *Harvard Business Review* article 'Disruptive Technologies: Catching the Wave', which argued that 'companies must give managers of disruptive innovation free

rein to realize the technology's full potential – even if it means ultimately killing the mainstream business. For the corporation to live, it must be willing to see business units die. If the corporation doesn't kill them off itself, competitors will.' See Bower, J. L., and Christensen, C. M., 'Disruptive Technologies: Catching the Wave', *Harvard Business Review* 73, no. 1 (January–February 1995): 43–53, online at https://hbr.org/1995/01/disruptive-technologies-catching-the-wave.

14 Sterman, J. D. (2006), 'Learning from evidence in a complex world', *American Journal of Public Health*, 96(3), 505–514. https://doi.org/10.2105/AJPH.2005.066043.

15 Ibid.

16 Harry Frankfurt, *On Bullshit* (Princeton, 2005), pp. 46 and 52. The Fall 1986 issue of *Raritan* magazine, vol. 6 no. 2, is online at https://raritanquarterly.rutgers.edu/issue-index/all-volumes-issues/volume-06/volume-06-number-2.

17 See '"This Week" Transcript: Donald Trump and Ben Carson', ABC News, 22 November 2015, online at https://abcnews.go.com/Politics/week-transcript-donald-trump-ben-carson/story?id=35336008.

18 Daniel Dennett, *From Bacteria to Bach and Back*, p. 215. The famous passage in *The Selfish Gene* in which Richard Dawkins coined the concept of 'memes' in 1976 argues that 'a new kind of replicator has recently emerged on this very planet. It is staring us in the face. It is still in its infancy, still drifting clumsily about in its primeval soup, but already it is achieving evolutionary change at a rate that leaves the old gene panting far behind. The new soup is the soup of human culture. We need a name for the new replicator, a noun which conveys the idea of a unit of cultural transmission, or a unit of imitation. "Mimeme" comes from a suitable Greek root, but I want a monosyllable that sounds a bit like "gene". I hope my classicist friends will forgive me if I abbreviate *mimeme* to *meme*.'

19 Jaron Lanier, *Ten Arguments for Deleting Your Social Media Accounts Right Now* (Vintage, 2019), p. 15.

20 Daniel Dennett, *From Bacteria to Bach and Back*, p. 220.

21 Ursula Le Guin, 'The Carrier Bag Theory of Fiction' (1986), collected in *Dancing at the Edge of the World* (Grove, 1997), pp. 166–7. This essay in particular draws on, and cites, Elizabeth Fisher's *Women's Creation* (McGraw-Hill, 1975) for its argument that the earliest cultural inventions were probably containers. I was reminded of Le Guin's essay, and its significance, by Siobhan Leddy's excellent piece 'We should all be reading

more Ursula Le Guin' of 28 August 2019 for *The Outline*, online at https://theoutline.com/post/7886/ursula-le-guin-carrier-bag-theory.
22 Jonathan Haidt, *The Righteous Mind* (Pantheon, 2012), p. 328.
23 Alasdair MacIntyre, *After Virtue: A Study in Moral Theory* (1981: 3rd edn., Bloomsbury, 2013), p. 250.

CHAPTER 7: VALUES AND ASSUMPTIONS: THE DELUSION OF NEUTRALITY

1 For an influential recent account of the nature of the existential risks and opportunities we now face, see Toby Ord, *The Precipice: Existential Risk and the Future of Humanity* (Bloomsbury, 2020). For some of the philosophical foundations of modern long-termist thinking, see part four of Derek Parfit's masterpiece *Reasons and Persons* (Clarendon Press, 1984) and its account of what we owe future generations.
2 In October 1985, Kranzberg gave an address at the Henry Ford Museum, Michigan, in which he proposed six 'laws' of technology as a summation of his core beliefs. The first reads: 'Technology is neither good nor bad; nor is it neutral.' The sixth is also particularly relevant to my arguments: 'Technology is a very human activity – and so is the history of technology.' For full details see Kranzberg, M. (1995). 'Technology and History: "Kranzberg's Laws."' *Bulletin of Science, Technology and Society*, 15(1), 5–13, https://doi.org/10.1177/027046769501500104.
3 Latour, B. (1992), 'Where Are the Missing Masses? The Sociology of a Few Mundane Artifacts', in W. E. Bijker and J. Law (eds.), *Shaping Technology/Building Society: Studies in Sociotechnical Change* (MIT Press, 1992), pp. 225–58.
4 A slogan used by the US National Rifle Association whose sentiments date back at least to the 1920s line from gun manufacturer Colt: 'It's not the gun. It's the man behind the gun.' For an in-depth discussion of its place and significance in American life, see Dennis A. Henigan, *"Guns Don't Kill People, People Kill People": And Other Myths About Guns and Gun Control* (Beacon, 2016).
5 In *The Senses Considered as Perceptual Systems* (Houghton and Mifflin, 1966).
6 J. J. Gibson, *The Ecological Approach to Visual Perception* (Houghton Mifflin Harcourt, 1979), ch. 8, 'The Theory of Affordances', p. 127.
7 Ibid., p. 130.
8 Kevin Kelly, *What Technology Wants* (Viking, 2010), pp. 15–16.

9 Once again, Thomas P. Hughes's theory of technological momentum is relevant here. See 'Technological Momentum' in Merritt Roe Smith and Leo Marx (eds.), *Does Technology Drive History? The Dilemma of Technological Determinism* (MIT Press, 1994).

10 For a fine essay exploring the hazards resulting from a sedentary lifestyle – and the possible alternatives – see the delightfully titled 'Arsebestos' by science fiction master Neal Stephenson, collected in *Some Remarks* (William Morrow, 2012). Sample warning: 'If you sit for any significant amount of time per day, it will kill you.'

11 Shannon Vallor, *Technology and the Virtues: A Philosophical Guide to a Future Worth Wanting* (OUP, 2016), pp. 2–3.

12 See – once again! – the thirteen essays collected in Merritt Roe Smith and Leo Marx's *Does Technology Drive History?*, which offer a subtle range of reflections upon technological influence – and, very broadly, conclude that it is both enormously significant while also being social, contingent and mutable.

13 The Borg first appear in the *Star Trek: The Next Generation* episode 'Q Who', Season 2, Episode 16; but they don't utter this line until their second appearance, in 'Best of Both Worlds, Part 1', Season 3, Episode 26.

14 L. M. Sacasas, 'Borg Complex: A Primer', 1 March 2013, The Frailest Thing blog, online at https://thefrailestthing.com/2013/03/01/borg-complex-a-primer/.

15 'Evernote CEO Phil Libin Interview: Evernote Business, Coal Mines And "The Nike Of The Well-Ordered Mind"', interview with Michael Rundle for HuffPost, 20 December 2012, online at https://www.huffingtonpost.co.uk/2012/12/20/evernote-ceo-phil-libin-interview-business-robotics-nike_n_2338111.html. This interview is highlighted by Sacasas in the previously cited piece on the Borg Complex.

16 For a fine account of this (and the corporate structure that made it all possible), see John Naughton, 'A moment's silence, please, for the death of Mark Zuckerberg's metaverse', *Guardian*, 13 May 2023, online at https://www.theguardian.com/technology/commentisfree/2023/may/13/death-of-mark-zuckerberg-metaverse-meta-facebook-virtual-reality-ai.

17 Brett Frischmann and Evan Selinger, *Re-Engineering Humanity* (CUP, 2018), p. 18.

18 Ibid.

19 Todd Feathers and Janus Rose, 'Students Are Rebelling Against Eye-Tracking Exam Surveillance Tools', Motherboard, 24 September 2020,

NOTES

online at https://www.vice.com/en/article/n7wxvd/students-are-rebelling-against-eye-tracking-exam-surveillance-tools.
20 For example, see Jason Kelley, 'Students Are Pushing Back Against Proctoring Surveillance Apps', Electronic Frontier Foundation, 25 September 2020, online at https://www.eff.org/deeplinks/2020/09/students-are-pushing-back-against-proctoring-surveillance-apps, as well as the wider advocacy of the EFF around this theme.
21 Sioux McKenna, 'Universities shouldn't use software to monitor online exams: here's why', *The Conversation*, 12 August 2022, online at https://theconversation.com/universities-shouldnt-use-software-to-monitor-online-exams-heres-why-188327.
22 Evan Greer and Evan Selinger, 'How facial recognition technology could change college campuses completely', MTV News, 28 February 2020, online at https://www.mtv.com/news/3158015/facial-recognition-technology-college-campuses/.
23 Brett Frischmann and Evan Selinger, *Re-Engineering Humanity*, p. 27.

CHAPTER 8: MYTHS AND WISH-FULFILMENT:
THE DELUSION OF MAGICAL THINKING

1 Joan Didion, *The Year of Magical Thinking* (Knopf, 2005), p. 35.
2 Ibid., p. 188.
3 Ibid., p. 206.
4 Simone de Beauvoir, *Les Belles Images* (Putnam, 1968), trans. Patrick O'Brian, p. 51.
5 Jacques Ellul, *The Technological Society* (Vintage, 1964), translated from the 1954 French edition by John Wilkinson, pp. 323–4. I am grateful to the philosopher L. M. Sacasas for introducing me to Ellul's work, and for the rich discussions of it that have appeared in his newsletter 'The Convivial Society'.
6 Ibid., p. 145.
7 See https://www.archives.gov/milestone-documents/tennessee-valley-authority-act – as the US National Archive notes in its commentary, 'the establishment of the TVA marked the first time that an agency was directed to address the total resource development needs of a major region'.
8 Evgeny Morozov, *To Save Everything, Click Here* (Allen Lane, 2013), pp. 5–6.

NOTES

9. 'Where do you want to go today' was, in fact, the title and main logo of Microsoft's 1994 global image advertising campaign.
10. Ivan Illich, *The Right to Useful Unemployment and Its Professional Enemies* (Boyars, 1978), p. 68.
11. Evgeny Morozov, *To Save Everything, Click Here*, p. 358.
12. Thomas Rid, *Rise of the Machines: the lost history of cybernetics* (Scribe, 2016), p. 2.
13. Ursula K. Le Guin, 'Myth and Archetype in Science Fiction' (1976), collected in *The Language of the Night*, ed. Susan Wood (Berkley, 1982), pp. 63–4.
14. Ursula K. Le Guin, 'The Beast in the Book', a talk given at the Conference on Literature and Ecology in Eugene, Oregon, June 2005, revised in 2014; collected in *Words Are My Matter: Writings on Life and Books* (Mariner Books, 2019), pp. 26 and 34.
15. Ursula K. Le Guin, 'Myth and Archetype in Science Fiction' (1976) in *The Language of the Night*, p. 65.
16. Book X, *The Metamorphoses of Ovid*, trans. Henry T. Riley (George Bell & Sons, 1893), online at https://www.gutenberg.org/files/26073/26073-h/26073-h.htm.
17. I have modernized a few spellings, but otherwise left the original text unchanged. The passage is from Book V, Canto II, Stanza XXVI of *The Faerie Queene*. 'Silver trye' refers to silver that has been 'tried' or purified, i.e. her feet are being likened to pure silver, to match her golden hands – not to mention the money her father took from honest knights, and with which she fruitlessly tried to bribe Artegal and Talus.
18. Cameron, J., et al. (1991), *Terminator 2: Judgment Day*, Carolco Pictures, Pacific Western Productions, Lightstorm Entertainment, Le Studio Canal+.
19. While I have no idea whether Ursula Le Guin watched or admired *Terminator 2*, her verdict on one of its writer/director James Cameron's other blockbusters is on record – and leads me to suspect that she may have viewed its violence as itself embodying a regrettable form of magical thinking. After several commentators noted that the plot of Cameron's 2009 film *Avatar* closely echoed Le Guin's 1976 novella *The Word for World is Forest*, she explicitly noted in her Introduction to the 2017 Library of America edition of *The Hainish Novels and Stories, Volume Two* that 'a high-budget, highly successful film resembled [*Word for World*] in so many ways that people have often assumed I had some part in making it. Since the film completely reverses the book's moral premise, presenting

the central and unsolved problem of the book, mass violence, as a solution, I'm glad I had nothing at all to do with it.'
20 Simone de Beauvoir would, I think, have approved. As she put it in *The Ethics of Ambiguity* (1948), 'To be free is not to have the power to do anything you like; it is to be able to surpass the given towards an open future.'
21 The other two 'laws' are: 'When a distinguished but elderly scientist states that something is possible, they are almost certainly right. When they state that something is impossible, they are very probably wrong' and 'The only way of discovering the limits of the possible is to venture a little way past them into the impossible.' Clarke added the footnote containing the third law to the 1973 edition of the 1962 essay in which the first two appeared, 'Hazards of Prophecy: The Failure of Imagination', having first coined the phrase in a 1968 letter to *Science* magazine. The letter itself was written in response to a letter from his friend Isaac Asimov invoking Clarke's original 1962 essay. For the letter, see Arthur C. Clarke, 'Clarke's Third Law on UFO's', *Science*, vol. 159 no. 3812, 19 January 1968: 255.
22 Arthur C. Clarke, *Profiles of the Future: an inquiry into the limits of the possible* (Bantam Books, 1962), p. 19.
23 H. Geiger and E. Marsden (1913), 'The laws of deflexion of α particles through large angles', *The London, Edinburgh, and Dublin Philosophical Magazine and Journal of Science*, 25:148, 604–623, doi:10.1080/14786440408634197.
24 Ernest Rutherford (1911), 'The Scattering of α and β Particles by Matter and the Structure of the Atom', *Philosophical Magazine*, Series 6. 21 (125): 669–688, doi:10.1080/14786440508637080. A scan of the original paper can be found online at https://personal.math.ubc.ca/~cass/rutherford/r.pdf.
25 His comments were reported in *Nature* following a lecture by Rutherford on 11 September that year – see F., A., 'Atomic Transmutation', *Nature* 132, 432–433 (1933), https://doi.org/10.1038/132432a0.
26 Arthur C. Clarke, *Profiles of the Future*, p. 14.
27 See Dan Gardner and Philip E. Tetlock, *Superforecasting: The Art and Science of Prediction* (Crown, 2015).
28 For the second of two 1932 papers reporting findings from this early accelerator, see J. D. Cockcroft and E. T. S. Walton, 'Experiments with High Velocity Positive Ions. II. The Disintegration of Elements by High Velocity Protons', *Proceedings of the Royal Society A*, vol. 137, pp. 229–242, 1932. Cockcroft and Walton would win the 1951 Nobel Prize in physics

for their work; Rutherford himself had won the Nobel Prize in chemistry in 1908 'for his investigations into the disintegration of the elements, and the chemistry of radioactive substances'.

29 The paper Jenkin, J. G. (2011), 'Atomic Energy is "Moonshine"': What did Rutherford *Really* Mean?'. *Phys. Perspect.* 13, 128–145, https://doi.org/10.1007/s00016-010-0038-1, makes the case that Rutherford deliberately underplayed the potentials of nuclear energy because he feared its catastrophic power; I'm not entirely convinced by this argument but it emphasizes that there's plenty of room for debate – and that Rutherford was (of course) more than aware of the vast amounts of energy that could in theory be unleashed by splitting nuclei.

30 From his 1987 essay 'What are scientific revolutions?' collected in Thomas Kuhn, *The Road Since Structure: Philosophical Essays, 1970–1993, with an Autobiographical Interview*, ed. James Conant and John Haugeland (University of Chicago, 2000). p. 16.

31 Ibid., p. 17.

32 Carlo Rovelli, *There Are Places in the World Where Rules Are Less Important Than Kindness* (Allen Lane, 2020), trans. Erica Segre and Simon Carnell, p. xx.

33 Sigmund Freud, *A General Introduction to Psychoanalysis* (Boni and Liveright, 1920), trans. G. Stanley Hall, pp. 246–7.

34 Ibid., p. 247.

35 Luciano Floridi, *The Fourth Revolution: How the Infosphere is Reshaping Human Reality* (OUP, 2014), pp. 90 and 93.

CHAPTER 9: TRICKERY AND INTELLECT: THE ANTHROPOMORPHIC DELUSION

1 Tom Standage, *The Mechanical Turk: The True Story of the Chess-Playing Machine That Fooled the World* (Penguin, 2003), pp. 217–18.

2 Ibid., p. 219.

3 Ibid., p. 224.

4 Carolyn Marvin, *When Old Technologies Were New: Thinking About Electric Communication in the Late Nineteenth Century* (Oxford University Press, 1988), p. 57.

5 Tom Standage, *The Mechanical Turk*, p. 220.

6 A. M. Turing, 'Computing Machinery and Intelligence', *Mind*, Volume LIX, Issue 236, October 1950, pp. 433–460, https://doi.org/10.1093/mind/LIX.236.433.

NOTES

7 Ibid.

8 For a thoughtful book exploring the Loenber Prize and the author's own participation in it, see Brian Christian's *The Most Human Human: What Artificial Intelligence Teaches Us About Being Alive* (Penguin, 2012).

9 E. Demchenko and V. Veselov, 'Who Fools Whom? The Great Mystification, or Methodological Issues on Making Fools of Human Beings', in Robert Epstein, Gary Roberts, Grace Beber (eds.), *Parsing the Turing Test: Philosophical and Methodological Issues in the Quest for the Thinking Computer* (Springer, 2009), p. 458.

10 Joseph Weizenbaum, *Computer Power and Human Reason* (W.H. Freeman, 1976), pp. 2–4. It's worth noting that this kind of interaction also suggests some important ethical questions around whether people have a right to know if they're interacting with a human or a machine – and the uses the content of their interactions may subsequently be put to. Among other things, 'Turing flags' have been discussed as a way of indicating you're dealing with an AI rather than a human. See, for example, Walsh, T., 'Turing's Red Flag', *Communications of the ACM*, July 2016, Vol. 59 No. 7, pp. 34–37, doi:10.1145/2838729, online at https://cacm.acm.org/magazines/2016/7/204019-turings-red-flag/fulltext, which makes the case for a 'Turing Red Flag law' intended to ensure that 'an autonomous system should be designed so that it is unlikely to be mistaken for anything besides an autonomous system, and should identify itself at the start of any interaction with another agent'.

11 Wikipedia's article on software easter eggs is a fine place to start if you want to dive down this particular rabbit hole: https://en.wikipedia.org/wiki/Easter_egg_(media). The term itself, as the article points out, originates with 1980s video games, and the game-like delight of hunting for secrets is a large part of the appeal.

12 Sherry Turkle, *Alone Together: Why We Expect More from Technology and Less from Each Other*, 3rd edition (Basic Books, 2017), p. 1.

13 For a fine discussion of the connections between Taylor's ideas and technology, see L. M. Sacasas's essay 'The Analog City and the Digital City', *The New Atlantis*, Winter 2020, online at https://www.thenewatlantis.com/publications/the-analog-city-and-the-digital-city. For a wider discussion featuring Taylor and others, see M. Meijer and H. De Vriese (eds.), *The Philosophy of Reenchantment* (Routledge, 2021), https://doi.org/10.4324/9780367823443

14 For Deep Mind's detailed account of the principles behind this last victory, and reflections upon the development of game-playing AIs to

that point, see the article by the AlphaStar team 'AlphaStar: Mastering the real-time strategy game StarCraft II', 24 January 2019, online at https://www.deepmind.com/blog/alphastar-mastering-the-real-time-strategy-game-starcraft-ii.
15 See https://ai.meta.com/research/cicero/.
16 An up-to-date list of the world's fastest computers is maintained online at https://www.top500.org/.
17 See Stuart Russell, *Human Compatible* (Penguin, 2019), p. 61.
18 Ray Kurzweil, *The Age of Spiritual Machines* (Viking, 1999), p. 5.
19 This interaction took place on 3 March 2023 via https://chat.openai.com.
20 For the original paper detailing the development of Transformers by researchers at Google, see Vaswani, Ashish; Shazeer, Noam; Parmar, Niki; Uszkoreit, Jakob; Jones, Llion; Gomez, Aidan N.; Kaiser, Lukasz; Polosukhin, Illia (5 December 2017). 'Attention Is All You Need'. https://arxiv.org/abs/1706.03762. Specifically, Pretrained Transformers deploy a combination of an initial unsupervised 'pretraining' phase in order to attain an underlying level of linguistic fluency, followed by a supervised training phase that fine-tunes this foundational model for a particular task or context.
21 Ibid.
22 See Dan Milmo and agency, 'Two US lawyers fined for submitting fake court citations from ChatGPT', *The Guardian*, 23 June 2023, online at https://www.theguardian.com/technology/2023/jun/23/two-us-lawyers-fined-submitting-fake-court-citations-chatgpt.
23 Daniel Dennett, 'The problem with counterfeit people', *The Atlantic*, 16 May 2023, online at https://www.theatlantic.com/technology/archive/2023/05/problem-counterfeit-people/674075/.
24 Rachel Metz, 'No, Google's AI is not sentient', CNN Business, 14 June 2022, online at https://edition.cnn.com/2022/06/13/tech/google-ai-not-sentient.
25 In the realm of material science, for example, see Gregory S. Doerk et al., 'Autonomous discovery of emergent morphologies in directed self-assembly of block copolymer blends', *Science Advances*, 9 (2), doi:10.1126/sciadv.add3687, while for an account of the influence of AlphaFold on research into fields such as drug discovery, enzyme development and antibiotic resistance, see https://oecd.ai/en/wonk/alphafold-ai-accelerate-scientific-discovery.
26 Emily M. Bender, 'On NYT Magazine on AI: Resist the Urge to be Impressed', Medium, 18 April 2022, online at https://medium.com/@

emilymenonbender/on-nyt-magazine-on-ai-resist-the-urge-to-be-impressed-3d92fd9a0edd.

27 For an influential and accessible paper that explains the limitations of 'understanding' on the part of Natural Language Processing, see Emily M. Bender and Alexander Koller (2020), 'Climbing towards NLU: On Meaning, Form, and Understanding in the Age of Data'. *Proceedings of the 58th Annual Meeting of the Association for Computational Linguistics*, pp. 5185–5198, online at https://aclanthology.org/2020.acl-main.463.pdf. Most famously, the paper uses the thought experiment of a hyper-intelligent deep sea octopus eavesdropping (via an underwater cable) on electronic communications between two humans. The octopus (O) learns to predict the patterns of language in these communications with immense sophistication, and can plausibly generate its own new patterns in response, such that it can successfully pose as one of the humans by hacking into the cable. It thus passes a weak version of the Turing test. But, having never encountered any of the surface-world objects under discussion, the ruse is revealed when one of A and B faces a bear attack and asks the octopus (thinking it is the other human) for suggestions on how to construct a weapon for self-defence. As the paper puts it: 'Solving a task like this requires the ability to map accurately between words and real-world entities (as well as reasoning and creative thinking). It is at this point that O would fail the Turing test, if A hadn't been eaten by the bear before noticing the deception. Having only form available as training data, O did not learn meaning. The language exchanged by A and B is a projection of their communicative intents through the meaning relation into linguistic forms. Without access to a means of hypothesizing and testing the underlying communicative intents, reconstructing them from the forms alone is hopeless, and O's language use will eventually diverge from the language use of an agent who can ground their language in coherent communicative intents.'

28 The statement was published by the Center for AI Safety on 30 May 2023, and is online at https://www.safe.ai/statement-on-ai-risk. It came off the back of a 22 March 2023 open letter, published via the Future of Life Institute, that called 'on all AI labs to immediately pause for at least 6 months the training of AI systems more powerful than GPT-4' in order to address some of the risks inherent to them: https://futureoflife.org/open-letter/pause-giant-ai-experiments/.

29 Brian Merchant, 'Afraid of AI? The startups selling it want you to be', *LA Times*, 31 March 2023, online at https://www.latimes.com/business/

technology/story/2023-03-31/column-afraid-of-ai-the-startups-selling-it-want-you-to-be.

30 For a powerful survey of just some of the work that has been done in this area, see Safiya Umoja Noble, *Algorithms of Oppression: How Search Engines Reinforce Racism* (NYU, 2018), and Cathy O'Neil, *Weapons of Math Destruction: How Big Data Increases Inequality and Threatens Democracy* (Crown, 2016).

31 For one of the most influential books to popularize this area, see Atul Gawande, *The Checklist Manifesto* (Metropolitan, 2009). And for a masterful account of the benefits that can flow from reducing unwanted variance created by human factors, see Cass R. Sunstein, Daniel Kahneman, and Olivier Sibony, *Noise: A Flaw in Human Judgement* (William Collins, 2021).

32 See 'QandA with John Tasioulas, AI2050 Senior Fellow', online at https://ai2050.schmidtfutures.com/community-perspective-john-tasioulas/.

33 Emily M. Bender, Timnit Gebru, Angelina McMillan-Major, and Shmargaret Shmitchell (a pseudonym used by Margaret Mitchell because her then employer, Google, wouldn't permit her to put her name to the paper), 'On the Dangers of Stochastic Parrots: Can Language Models Be Too Big?', 1 March 2021, *Proceedings of the 2021 ACM Conference on Fairness, Accountability, and Transparency* (FAccT '21). Association for Computing Machinery, New York, NY, USA, 610–623, https://doi.org/10.1145/3442188.3445922.

34 Emily M. Bender, 'On NYT Magazine on AI: Resist the Urge to be Impressed'.

35 Although versions of this sentiment have been used countless times to describe online business models, one of its first instances was a 1973 video by the artists Richard Serra and Carlota Fay Schoolman titled 'Television Delivers People' which displayed the following sentences onscreen: 'It is the consumer who is consumed. You are the product of t.v. You are delivered to the advertiser who is the customer. He consumes you. The viewer is not responsible for programming . . . You are the end product.' See the Quote Investigator website https://quoteinvestigator.com/2017/07/16/product/.

36 David Weinberger, 'Learn from machine learning', *Aeon*, 15 November 2021, online at https://aeon.co/essays/our-world-is-a-black-box-predictable-but-not-understandable. See also his fine 2019 book *Everyday*

Chaos (Harvard Business Review Press) for an in-depth exploration of these and related ideas.

37 Ted Chiang, 'ChatGPT is a blurry JPEG of the web', 9 February 2023, *The New Yorker*, online at https://www.newyorker.com/tech/annals-of-technology/chatgpt-is-a-blurry-jpeg-of-the-web.

38 See Shumailov, I., Shumaylov, Z., Zhao, Y., Gal, Y., Papernot, N., and Anderson, R. 'The Curse of Recursion: Training on Generated Data Makes Models Forget', https://doi.org/10.48550/arXiv.2305.17493.

39 See 'Exclusive: OpenAI Used Kenyan Workers on Less Than $2 Per Hour to Make ChatGPT Less Toxic', *Time*, 18 January 2023, online at https://time.com/6247678/openai-chatgpt-kenya-workers/.

CHAPTER 10: SUPERINTELLIGENCE AND DOUBT: THE DELUSION OF MACHINE PERFECTION

1 'Highway Accident Report: Collision Between Vehicle Controlled by Developmental Automated Driving System and Pedestrian, Tempe, Arizona, March 18, 2018', National Transportation Safety Board Accident Report NTSB/HAR-19/03 PB2019-101402, adopted 19 November 2019, online at https://data.ntsb.gov/Docket/?NTSBNumber=HWY18MH010.

2 For a detailed account of events from Vasquez's perspective, see Lauren Smiley, '"I'm the Operator": The Aftermath of a Self-Driving Tragedy', *Wired*, 8 March 2022, online at https://www.wired.com/story/uber-self-driving-car-fatal-crash/. Here is the *Wired* article's account of her streaming activity at the time of the crash: 'Last summer, her two new lawyers loosed Vasquez's defense in a pretrial legal filing: Yes, she was streaming *The Voice* on Hulu, the defense wrote – but she wasn't watching it; she was listening to it. And that was something operators were allowed to do.' In July 2023, Vasquez pled guilty to one count of endangerment and was sentenced to three years' probation: see https://www.theguardian.com/technology/2023/aug/01/uber-self-driving-arizona-deadly-crash

3 Importantly, the original thought experiment was intended to illuminate the ethics of abortion rather than any literal issues around vehicles; the 'trolley' scenario was intended to illustrate the difference between the 'positive' duty to help someone versus the 'negative' duty to avoid harming them. See Foot, Philippa (1967). 'The Problem of Abortion and the Doctrine of the Double Effect.' *Oxford Review* 5:5–15, online at

https://philarchive.org/archive/FOOTPO-2v1. And for a highly entertaining book-length exploration of the thought experiment's legacy, see David Edmonds, *Would You Kill the Fat Man? The Trolley Problem and What Your Answer Tells Us about Right and Wrong* (Princeton, 2013).

4 'Highway Accident Report: Collision Between Vehicle Controlled by Developmental Automated Driving System and Pedestrian, Tempe, Arizona, March 18, 2018', National Transportation Safety Board Accident Report NTSB/HAR-19/03 PB2019-101402, adopted 19 November 2019, p. 59, online at https://data.ntsb.gov/Docket/?NTSBNumber=HWY18MH010.

5 Elish, Madeleine Clare, 'Moral Crumple Zones: Cautionary Tales in Human-Robot Interaction' (pre-print) (March 1, 2019). *Engaging Science, Technology, and Society* (pre-print), available at SSRN: https://ssrn.com/abstract=2757236 or http://dx.doi.org/10.2139/ssrn.2757236

6 'Highway Accident Report', pp. 16–17.

7 Ibid., p. 60.

8 Jack Stilgoe, *Who's Driving Innovation?* (Palgrave Macmillan, 2019), p. 1.

9 Ibid., pp. 21–22. The 'technological sublime' is a specific reference to David Nye's 1994 book *American Technological Sublime*, a rich historical and political study of the place of technology in American self-conceptions.

10 For a useful brief introduction to this area, and some practical guidance, see 'The seven tenets of human-centred design', online at https://www.designcouncil.org.uk/our-work/news-opinion/seven-tenets-human-centred-design/.

11 For the latest Society of Automotive Engineers guidance, see 'Taxonomy and Definitions for Terms Related to Driving Automation Systems for On-Road Motor Vehicles J3016_202104' online at https://www.sae.org/standards/content/j3016_202104/, and for details of Mercedes-Benz's certification for level three automation see ' Certification for SAE Level 3 system for U.S. market', 26 January 2023, online at https://group.mercedes-benz.com/innovation/product-innovation/autonomous-driving/drive-pilot-nevada.html

12 See https://waymo.com/waymo-driver/ for the latest details.

13 Cory Doctorow, 'Pluralistic: VW wouldn't locate kidnapped child because his mother wasn't paying for find-my-car subscription', 28 February 2023, online at https://pluralistic.net/2023/02/28/kinderwagen/#worst-timeline. In a 7 March 2023 press release, VW subsequently announced that it would be offering its 'connected vehicle emergency

services at no additional cost for five years for most model year 2020 to 2023 vehicles' and that it 'must and will do better for everyone that trusts our brand and for the law enforcement officials tasked with protecting us'. See https://media.vw.com/en-us/releases/1733.

14 Nicholas Carr, *The Glass Cage: Automation and Us* (Norton, 2014), p. 210.
15 See Sherry Turkle, *Alone Together: Why We Expect More from Technology and Less from Each Other*, 3rd edition (Basic Books, 2017).
16 McCarthy, John; Minsky, Marvin; Rochester, Nathan; Shannon, Claude, 'A Proposal for the Dartmouth Summer Research Project on Artificial Intelligence', 31 August 1955, online at http://www-formal.stanford.edu/jmc/history/dartmouth/dartmouth.html.
17 For the original DeepMind paper, see Vlad Mnih, Koray Kavukcuoglu, David Silver, Alex Graves, Ioannis Antonoglou, Daan Wierstra, and Martin Riedmiller. 'Playing atari with deep reinforcement learning.' arXiv preprint arXiv:1312.5602 (2013) online at https://www.deepmind.com/publications/playing-atari-with-deep-reinforcement-learning.
18 For an accessible overview of Marcus's recent position, see his essay 'Deep Learning Is Hitting a Wall' in *Nautilus* magazine, 10 March 2022, online at https://nautil.us/deep-learning-is-hitting-a-wall-238440/, and in particular its argument that 'deep learning systems are black boxes; we can look at their inputs, and their outputs, but we have a lot of trouble peering inside. We don't know exactly why they make the decisions they do, and often don't know what to do about them (except to gather more data) if they come up with the wrong answers. This makes them inherently unwieldy and uninterpretable, and in many ways unsuited for "augmented cognition" in conjunction with humans. Hybrids that allow us to connect the learning prowess of deep learning, with the explicit, semantic richness of symbols, could be transformative.'
19 Gary Marcus and Ernest Davis, *Rebooting AI: Building Artificial Intelligence We Can Trust* (Pantheon, 2019), pp. 177–78.
20 Stuart Russell, *Human Compatible* (Penguin, 2019), p. 10.
21 Ibid., p. 11.
22 For the proof of checkers' solution, see Jonathan Schaeffer, Neil Burch, Yngvi Björnsson, Akihiro Kishimoto, Martin Müller, Robert Lake, Paul Lu, and Steve Sutphen, 'Checkers is solved', *Science* 317, no. 5844 (2007): 1518–1522, doi:10.1126/science.1144079. For a recent estimate of the number of moves in chess (without promotions), see Stefan Steinerberger (2015), 'On the number of positions in chess without promotion', *International Journal of Game Theory* 44, 761–767, https://doi.

org/10.1007/s00182-014-0453-7, and for the classic paper by Claude Shannon on this topic, see Claude E. Shannon (1950), 'Programming a computer for playing chess', *Philosophical Magazine* 41: 314, online at https://vision.unipv.it/IA1/ProgrammingaComputerforPlayingChess.pdf.

23 Nick Bostrom (2003), 'Ethical issues in advanced artificial intelligence', *Science fiction and philosophy: from time travel to superintelligence*, 277, 284. Full text online at https://nickbostrom.com/ethics/ai.html.

24 'Nick Bostrom on artificial intelligence', 8 September 2014, interview for theOUPblog,onlineathttps://blog.oup.com/2014/09/interview-nick-bostrom-superintelligence/.

25 For an eloquent, in-depth exploration of debates around AIs' ultimate values and purposes, and what it might mean to make these compatible with human thriving, see Brian Christian, *The Alignment Problem: Machine Learning and Human Values* (Atlantic Books, 2021).

26 See, for example, Rebecca Ackermann's account in 'Inside effective altruism, where the far future counts a lot more than the present', *MIT Technology Review*, 17 October 2022, online at https://www.technologyreview.com/2022/10/17/1060967/effective-altruism-growth/.

27 Thomas Frank, 'Home of the Whopper', *Harper's Magazine*, November 2013, online at https://harpers.org/archive/2013/11/home-of-the-whopper.

28 Unfortunately, none of the suggestions in this paragraph are purely hypothetical. See, for example, the Vice Motherboard blog's reporting into the use by San Francisco police of recordings from autonomous vehicles as a form of surveillance: Aaron Gordon, 'San Francisco Police Are Using Driverless Cars as Mobile Surveillance Cameras', Motherboard, 11 May 2022, online at https://www.vice.com/en/article/v7dw8x/san-francisco-police-are-using-driverless-cars-as-mobile-surveillance-cameras.

CHAPTER II: TOWARDS A NEW ETHICS OF TECHNOLOGY: THE DELUSION OF DIVINE DATA

1 For a detailed account of epiphenomenalism as a philosophy of mind, see the Stanford Encyclopedia of Philosophy online at https://plato.stanford.edu/entries/epiphenomenalism/#:~:text=Epiphenomenalism.

2 The lecture was published later that year in *Psychology Review*. See Watson, J. B. (1913), 'Psychology as the behaviorist views it', *Psychological Review*, 20(2), 158–177, https://doi.org/10.1037/h0074428.

NOTES

3 Hawking was speaking at Google's Zeitgeist Conference in May 2011 about his recent book *The Grand Design*; for a video and transcript of the talk, see https://www.youtube.com/watch?v=pdLdA8E1Oao.
4 For example, see Matthew Reisz, 'Is philosophy dead?', *Times Higher Education*, 22 February 2015, online at https://www.timeshigher education.com/news/is-philosophy-dead/2018686.article, which details just a few of the many responses made by philosophers at the time.
5 Rani Lill Anjum and Stephen Mumford, *Causation in Science and the Methods of Scientific Discovery* (OUP, 2018), p. 6.
6 Ibid., p. 7.
7 Chris Anderson, 'The End of Theory: The Data Deluge Makes the Scientific Method Obsolete', *Wired*, 23 June 2008, online at https://www.wired.com/2008/06/pb-theory/.
8 Mary Midgley, *What is Philosophy For?* (Bloomsbury Academic, 2018), digital edition, locations 816 and 876. Popper's phrase 'promissory materialism' embodies a withering critique of the view that because materialism is the only way anything can ever be explained, it will in due course explain everything we don't currently understand. Popper and the neurophysiologist John Eccles coined it in their 1977 book *The Self and its Brain*, noting that in the case of such arguments: 'No attempt is made to resolve the difficulties of materialism by argument. No alternatives to materialism are even considered. Thus it appears that there is, rationally, not more of interest to be found in the thesis of promissory materialism than, let us say, in the thesis that one day we shall abolish cats or elephants by ceasing to talk about them . . . For all the physicalist offers is, as it were, a cheque drawn against his future prospects, and based on the hope that a theory will be developed one day which solves his problems for him; the hope, in short, that something will turn up.' (Routledge, 1984), pp. 97–8.
9 Shannon Vallor, *Technology and the Virtues: A Philosophical Guide to a Future Worth Wanting* (OUP, 2016), pp. 5–6.
10 Kant first sets out the Categorial Imperative in his 1785 *Groundwork of the Metaphysics of Morals*, with perhaps its most famous modern English-language form provided by James Ellington's 1981 translation: 'Act only according to that maxim whereby you can at the same time will that it should become a universal law', *Grounding for the Metaphysics of Morals. With On a Supposed Right to Lie Because of Philanthropic Concerns* (Hackett, 1993 edition), p. 30.
11 Shannon Vallor, *Technology and the Virtues*, p. 7.

NOTES

12 As I've suggested, there is a great deal to be learned from its emphasis on purposes, states of mind and responsibilities – not least in the extension of these properties to informational artefacts and systems. For just one example, see Andreas Spahn (2020), 'Digital Objects, Digital Subjects and Digital Societies: Deontology in the Age of Digitalization', *Information*; 11(4):228, https://doi.org/10.3390/info11040228, which 're-examines the three categories "subject", "object" and "intersubjectivity" . . . and suggests deontological guidelines for digital objects, digital subjects and a digitally mediated intersubjectivity, based on a re-examination of the requirements of epistemic, motivational and deliberational rationalism.'
13 For a thorough treatment of this important and influential contemporary ethical system, see Luciano Floridi, *The Ethics of Information* (OUP, 2013).
14 Jobin, A., Ienca, M. and Vayena, E. (2019), 'The global landscape of AI ethics guidelines', *Nature Machine Intelligence* 1, 389–399, https://doi.org/10.1038/s42256-019-0088-2.
15 See Peter Singer, *The Life You Can Save: Acting Now to End World Poverty* (Picador, 2010).
16 J. J. C. Smart and Bernard Williams, *Utilitarianism: For and Against* (CUP, 1973), pp. 149–150.
17 See Toby Ord's *The Precipice* (Bloomsbury, 2020), William MacAskill's *What We Owe the Future* (Oneworld, 2022), and the fourth section of Derek Parfit's *Reasons and Persons* (Clarendon Press, 1984), about future generations.
18 See also Nick Bostrom, *Superintelligence: Paths, Dangers, Strategies* (OUP, 2014). I don't share Bostrom's assumptions, but his is certainly an influential perspective. As his book concludes (p. 260): 'Through the fog of everyday trivialities, we can perceive – if but dimly – the essential task of our age . . . one that presents as our principal moral priority (at least from an impersonal and secular perspective) the reduction of existential risk and the attainment of a civilizational trajectory that leads to a compassionate and jubilant use of humanity's cosmic endowment.' Perhaps above all, I worry about this invocation of an 'impersonal and secular perspective' as the lens through which humanity's and technology's future should be viewed. Ironically, this seems likely to me to fail the most significant of consequentialist tests in the narrowness of its account of our values, priorities, motivations and self-understandings, and thus in its ability to model and influence these. To borrow Bernard

Williams's phrase, there are 'too few thoughts and feelings to match the world as it really is'.

19 Kate Crawford, *The Atlas of AI: Power, Politics, and the Planetary Costs of Artificial Intelligence* (Yale University Press, 2021), p. 8.

20 Edith Hall, *Aristotle's Way* (Vintage, 2018), digital edition, location 423.

21 Julian Baggini, *The Godless Gospel: Was Jesus A Great Moral Teacher?* (Granta, 2020), pp. 33-34.

22 Joy Buolamwini, 'How I'm fighting bias in algorithms', TEDxBeaconStreet, November 2016, transcript online at https://www.ted.com/talks/joy_buolamwini_how_i_m_fighting_bias_in_algorithms/transcript.

23 See Megan Rose Dickey, 'Twitter and Zoom's algorithmic bias issues', Tech Crunch, 21 September 2020, online at https://techcrunch.com/2020/09/21/twitter-and-zoom-algorithmic-bias-issues/; and Alex Hern, 'Twitter apologises for "racist" image-cropping algorithm', *The Guardian*, 21 September 2020, online at https://www.theguardian.com/technology/2020/sep/21/twitter-apologises-for-racist-image-cropping-algorithm.

24 Joy Buolamwini, 'How I'm fighting bias in algorithms'.

25 As just one example of important ongoing work in this area, Black in AI, a global community that has grown to over 5,000 members, was founded in 2017 by the computer scientists Rediet Abebe and Timnit Gebru as 'a place for sharing ideas, fostering collaborations and discussing initiatives to increase the presence of Black people in the field of Artificial Intelligence.' See https://blackinai.github.io/#/.

26 See https://www.ajl.org/take-action.

27 Alex Hanna, Emily Denton, Razvan Amironesei, Andrew Smart and Hilary Nicole, 'Lines of Sight', *Logic(s)*, Issue 12, 20 December 2020, online at https://logicmag.io/commons/lines-of-sight. See also this paper: Emily Denton, Alex Hanna, Razvan Amironesei, Andrew Smart and Hilary Nicole (2021), 'On the genealogy of machine learning datasets: A critical history of ImageNet', *Big Data and Society*, https://doi.org/10.1177/20539517211035955.

28 Alasdair MacIntyre, *Dependent Rational Animals: Why Human Beings Need the Virtues* (Open Court, 2009), digital edition, location 2183.

29 Ibid., location 46.

30 Virginia Held, *The Ethics of Care: Personal, Political, and Global* (OUP, 1997), p. 73. I am indebted to the philosopher David Weinberger for introducing me both to care ethics as a discipline and to this particular

line from Held's work, which he cites in his own unpublished essay 'A Caring Ethics from Uncaring AI' – itself a brilliantly concise meta-ethical exposition of the unique qualities of care ethics in the context of technology. The foundational work of figures like Carol Gilligan and Nel Noddings is also significant in this field.

31 Elaine Castillo, *How to Read Now: Essays* (Atlantic Books, 2022), digital edition, location 305.

32 Carissa Véliz, *Privacy is Power* (Bantam Press, 2020), p. 201.

CHAPTER 12: DEATH AND LIFE: THE DELUSION OF PERPETUAL PROGRESS

1 See https://www.alcor.org/. All quotations last retrieved in January 2023, except as stated.

2 Ibid.

3 For an interesting survey of the uses and early potentials of cryonics, see Tae Hoon Jang, Sung Choel Park, Ji Hyun Yang, Jung Yoon Kim, Jae Hong Seok, Ui Seo Park, Chang Won Choi, Sung Ryul Lee and Jin Han (2017), 'Cryopreservation and its clinical applications', *Integrative Medicine Research*, 6(1), 12–18. https://doi.org/10.1016/j.imr.2016.12.001.

4 See https://www.alcor.org/what-is-cryonics/ (retrieved 20 May 2022).

5 For an updated version of the influential Transhumanism FAQ website first developed in the mid-1990s, see https://whatistranshumanism.org/ while for details about the history and influence of the original FAQ, see https://www.humanityplus.org/transhumanist-faq. As for the origins of the term itself, its oldest form can in fact be traced to the third volume of the *Divine Comedy*, the *Paradiso*, by the fourteenth-century Italian poet Dante Alighieri. The poem is narrated in the first person and takes the form of a pilgrimage. Guided by the spirit of the Roman poet Virgil, Dante has seen the worst of humanity suffering exemplary punishment, followed by visions of the battle between vice and virtue. Now he finds himself in the company of his beloved, Beatrice, whose spirit lifts him towards the celestial spheres. In wonder, he coins a new word to capture something of that experience: *trasumana*, 'to pass beyond the human'. Like much of the poetry of the Middle Ages, the *Divine Comedy* is a work in which nothing is merely literal. Throughout, it enacts the belief that ultimate questions can only be approached through a mixture of myth, allegory and visionary witness. By contrast, the English word 'transhumanism' was popularized in a 1957 essay by the biologist Julian

Huxley as a rallying cry for an antithetical vision of transcendence, a literal, bodily species of self-enhancement brought about solely by human endeavour: 'We are already justified in the conviction that human life as we know it in history is a wretched makeshift, rooted in ignorance; and that it could be transcended by a state of existence based on the illumination of knowledge and comprehension, just as our modern control of physical nature based on science transcends the tentative fumblings of our ancestors . . . We need a name for this new belief. Perhaps transhumanism will serve . . .' Unlike Dante's spiritual ecstasy, Huxley's is a resolutely secular vision. Through science and technology, he suggests, humanity is becoming the agent of its own transformation. Guided by rational illumination, we will be rescued from history's ignorant purgatory, our innermost natures known and controlled as surely as the material world. For the full context of the original use, see Dante Alighieri, *Paradiso*, Book I, lines 70–73: 'Passing beyond the human cannot be / worded; let Glaucus serve as simile – / until grace grant you the experience.' This translation is by Allen Mandelbaum, while Dante's original reads: '*Trasumanar significar per verba / non si poria; però l'essemplo basti / a cui esperïenza grazia serba.*' A complete parallel text of the original Italian and both Mandelbaum's and Longfellow's translations can be found online at the superb Digital Dante website https://digitaldante.columbia.edu/dante/divine-comedy/. I am indebted to conversations with the author Mark Vernon, and to his book *Dante's Divine Comedy: A Guide for the Spiritual Journey* (Angelico Press, 2021), for drawing my attention to Dante's relevance in the context of transhumanism. Huxley's essay was originally published in *New Bottles for New Wine* (Chatto and Windus, 1957), pp. 13–17.

6 Here's Nick Bostrom, for example, explaining why he thinks it's plausible that our entire universe is a simulation in his paper 'Are You Living in a Computer Simulation?', first published in *Philosophical Quarterly* (2003) Vol. 53, No. 211, pp. 243–255. 'Many works of science fiction as well as some forecasts by serious technologists and futurologists predict that enormous amounts of computing power will be available in the future. Let us suppose for a moment that these predictions are correct. One thing that later generations might do with their super-powerful computers is run detailed simulations of their forebears or of people like their forebears. Because their computers would be so powerful, they could run a great many such simulations. Suppose that these simulated people are conscious . . . Then it could be the case that the vast majority

of minds like ours do not belong to the original race but rather to people simulated by the advanced descendants of an original race.' Bostrom's argument is subtler than its conclusion might suggest. If, he suggests, future civilizations become capable of creating simulations within which conscious entities exist, then it's plausible that we live inside such a simulation. This is because a civilization capable of creating one such simulation would also, presumably, be capable of creating millions of them. By contrast, a chain of evolutionary events playing out on a particular planet is a one-off scenario. The theory that we evolved from scratch on such a planet is thus, according to Bostrom's line of reasoning, inherently less plausible than the theory that we exist inside one among millions of parallel computational contexts. In order to assess this argument, we need to weigh the plausibility of the 'if' statement with which it begins. If it is true that billions of simulated universes like ours can in principle exist, then it is indeed plausible that ours is one of them. But if the evolution of intelligent life is common, while complex simulations are rare, evolution starts to look more likely. If, meanwhile, it is true that a universe like ours (and, in particular, one that contains conscious life) cannot be simulated, then we cannot live in a simulation. How might we decide between these options? Despite the air of statistical authority it's possible to give various speculations; the answer is that – short of waiting for alien contact or technological apotheosis – there's little we can do when it comes to differentiating between these fundamentals. If, moreover, it is true both that a perfect simulation is indistinguishable from a non-simulation and that its 'outside' can never be accessed from its 'inside', it makes no material difference whether we're actually inside a simulation, no more than it matters whether we are actually dreaming, or fictional, or the abandoned children of an absent god. The world is what it is: only our stories have changed. See https://www.simulation-argument.com/simulation for the published paper.
7 Online at https://whatistranshumanism.org/#what-is-a-posthuman.
8 John Gray, *The Immortalization Commission: Science and the Strange Quest to Cheat Death* (Allen Lane, 2011), p. 216. Process theology itself is primarily a theological extension of the philosophy of Alfred North Whitehead (1861–1947), who emphasized the significance of 'becoming' over time, as opposed to the classical metaphysical emphasis upon being and essence.
9 Ibid., pp. 216–17.
10 Ibid., p. 221.

NOTES

11 Commentary by Claire Barliant with Nat Trotman on *This Progress* for the Guggenheim Collection Online at https://www.guggenheim.org/artwork/22502.
12 Samuel Scheffler, *Death and the Afterlife* (OUP, 2013), p. 18.
13 Ibid., p. 45.
14 Roman Krznaric, *The Good Ancestor: How to Think Long Term in a Short-Term World* (WH Allen, 2021), p. 202. Krznaric's summary diagram of the 'tug of war for time' can be found on page 12. Another fine book exploring the nature of long-term thinking in less prescriptive terms is *The Long View* (Wildfire, 2023) by Richard Fisher.

AND FINALLY

1 Andrew Solomon, *Far from the Tree: Parents, Children and the Search for Identity* (Vintage Books, 2014), p. 1.
2 W. H. Auden, 'September 1, 1939'. First published in *The New Republic*, 18 October 1939, then collected in its original form in *Another Time* (Random House, 1940).
3 It was read by Scott Simon on *Weekend Edition* on 15 September 2001; the programme is archived online at https://www.npr.org/templates/story/story.php?storyId=1129485.
4 The anthology in question was *The Poetry of the Thirties* (Penguin, 1964) and the full note, which referred to 'September 1, 1939' and four other early poems, read 'Mr. W. H. Auden considers these five poems to be trash which he is ashamed to have written'.
5 Andrew Solomon, *Far from the Tree*, p. 1.

Index

abacuses 62
accidental revelation, learning from 108–10, 118, 201
accountability 185, 186, 197, 201, 212, 231
Acheulean toolmaking industry 29–31
active/passive distinction 132–4, 137–8
Adams, Douglas 1, 2, 126, 202
adaptation 34–7, 48, 54–5, 74, 97, 101, 106, 113, 249–50
Aeon (magazine) 40, 188
affordances 124–6, 128, 132, 135, 137–8, 198, 201–2, 212, 217, 240, 254
Africa 27–8, 30–3, 38
    North 13
    sub-Saharan 27
agency 3, 29, 115, 117–18, 146, 191, 199, 201
    letting go of 212
aggression 47, 48, 49
AGI *see* Artificial General Intelligence
agriculture 14–18
AI *see* Artificial Intelligence
aircraft, combat 104
Al-Fārābī 228
Alcor 241–2
Alexa 172
algorithmic age 191
algorithmic bias 230
algorithmic data processing, normalization 137
Algorithmic Justice League 231

algorithms 115, 134, 136, 177–9, 182, 185, 204–5
alphabets 64–5
AlphaGo 174
AlphaStar 174
AlphaZero 174, 181–2
Amazon 172, 190–1
American Journal of Public Health 109
American Philosophical Association 233
American Psychological Association 215–16
Americas 27
Amironesei, Razvan 232
ancestors 9, 17–18, 25–8, 31, 37–9, 43, 45, 48, 50–4, 57, 71, 74, 97, 101
    see also hominins; hunter-gatherers
Ancient China 15
Ancient Egypt 15, 64–5
    Early Dynastic period 15
Ancient Greece 64–5
Anderson, Chris 218, 220, 221
anger, righteous 237–8
Anjum, Rani Lill 217–18
Anthropocene (Holocene) 40–1
anthropomorphic delusion 161–91, 253
anthropomorphism 180–7, 189, 253
antimony 11
ants 55, 57–8
apes 27, 28, 33, 43, 46–7, 55
Aphrodite 148
Apple 20, 105, 172, 174

INDEX

archaeology 16–18, 29, 52–4
architecture 16
Aristotle 155–6, 227–8, 237–8
art 69–70
Artegal, Sir 149–50
Arthur, W. Brian 10, 30
artificial artificial intelligence 190–1
Artificial General Intelligence (AGI), human-like nature 204
Artificial Intelligence (AI) 136, 150–2, 171–91
   accountability 231
   and consciousness 180–1
   costs of 186
   equitable 231
   ethics of 209, 223–4, 226–7, 231–2
   as existential risk to humanity 182–4, 186
   fairness of 232
   Generative 179
   and 'idealized machines' 198
   objectives of 207–11, 213
   and reality 177–81, 188
   rethinking 202–6
   understanding what we want from 187–91
   see also autonomous driving systems; chatbots
artisan products 15
Asia 27
   central 39
   Southeast 27, 32, 34, 38–9
assembly lines 211
assumptions 233, 253
   embedded in technology 122–3, 125, 127, 132–8
   as final and literal truth 156
Atari 205
Athens 68–9
atomic bombs 153, 183
atomic nucleus 153
attention 93–6, 100, 114
   and disinformation 111
   elasticity of 93
   ethics of 93–6, 115
      moral elements 94–5
   selectivity 93
   tracking 134
attentional engineering 95
Auden, W. H. 256
audiences 70
Australian Aboriginals 37
Australopithecus 27–8, 44–5
automation 171, 184, 188–9, 191, 198–9, 201–2, 206–7, 211–13, 230, 253
automatons 161–4, 166–7, 176, 190–1
autonomous driving systems 193–8, 199–202, 212–13
autonomy 133
   attribution to machines 164

baby-carrying slings 45
bacteria
   photosynthetic 36
   symbiotic 12
Baggini, Julian 91–2, 229
beliefs, magical 139–40
Bender, Emily M. 182, 186–7
bereavement 139–40
biases 103–8, 115–16, 231, 232, 253
Bible 10, 13
big data 221
Bildungsroman 159
bipedalism 44–5
birds 58
birth 255
blame 140
boats 99–104
body 59, 62, 71, 76–8, 180, 234
bonobos 27, 46, 55
book of Genesis 32
books 61–2, 71
Bostrom, Nick 209–10
'bottom up' processing 83–4, 86, 205–6
bow and arrow 128

INDEX

brain 71, 76–8, 79
  development 44
  literate 64
  manipulation 115
  and the mind 59
  and perception 83–6
  as prediction machine 86
  and the self 92
  size 45, 51
  stone age 51–2, 54, 98
Breakout (game) 205
Brisbane, Arthur 165
Bronze Age 15
brutality, delusion of 43–58, 252
Buddhism 11, 228
bullshit, overcoming 111–18
Buolamwini, Joy 229–31

Cameron, James 150
carbon dioxide levels 40
care 235, 249
  and attention 94
  intergenerational 58
  mutual 46–7, 54–5, 58, 233
care ethics 235
Carr, Nicholas 201
cars 122
  autonomous 193–8, 199–202, 212–13
Cartesian Theatre 85
Caselle, Piedmont 13
Castillo, Elaine 235–6
castration 46
categorical imperative 222
category errors 121, 156, 224, 233
cathedral thinking 251
Catherine the Great 162
cattle 14
cause and effect 15–16, 26, 102, 105, 215
cave art 39
  finger paintings 53–4
cephalopods 9
ceremonies, boat naming/launching 100

certainty 209, 245
  illusion of 93
  unwarranted 157
  see also uncertainty
cetaceans 9, 58
Chalmers, David 59, 62, 71–3, 78
change
  human capacity for 44, 74, 239
  social 230
chaos 246
Chartier, Émile-Auguste 100
chatbots 169, 170–2, 180–1
ChatGPT 177–9, 182, 191
checkers (game) 208
Chengdu 11
chess 161–4, 166–7, 173–8, 182, 189, 198–9, 208
Chiang, Ted 189
chickpeas 14
childcare, communal 46, 47–9, 58
childrearing, achieving immortality through 255–7
children/childhood 43–54, 56–8, 63, 73–5, 77, 97, 172–3, 233–5, 248, 255–7
chimps 27, 55
China
  paper-making 13
  wood block printing 11
Choe Yun-ui 11, 12
Christianity, Thomist 228
Cicero AI 174
citizenship, in the digital age 235, 237
city-states 16
civilization
  and agriculture 14–18
  emergence of 15–16
Clark, Andy 59, 62, 71–3, 78
Clarke, Arthur C. 152–5, 166
Clarke's third law 152–4
climate change 14–15, 33, 39–42
climate tipping points 42
Cobenzl, Count Ludwig von 161–2

INDEX

coercion 128
cognition 71–3
  embodied nature 76
coherence 76
coin minting 11
collaboration 17, 31, 33, 46–7, 55, 97, 190, 250
collective unconscious 62
communication 97
comparison-making, social 55–6
compassion 5, 49, 55, 240–1, 250
competence, superhuman 204–5
competition 47, 252
complexity 14–16, 95, 107–10, 129, 184, 202, 226, 232
  and the delusion of comprehension 253
  embracing 254
  exponential increase 14, 15, 18, 21, 220–1
  human 203
  of life on earth 250
  of modernity 236
  as norm 93
  of reality 188
  and simplicity 93
  social 16, 33
  of technology 14–15, 18, 20, 101, 202, 220–1, 224, 225, 253
comprehension, delusion of 97–118
computer simulations, existence as 243
computers 174–6
  chess-playing 167, 173–6
  complexity 20
  evolution 20–2
  supercomputers 152, 174
Conference on Literature and Ecology 2005 147–8
confirmation bias 115–16
conflict 58, 252
  see also delusion of brutality
Confucianism 228
consciousness 62
  as controlled hallucination 83–5, 91–2, 159, 253
  and death 242
  embodied nature 76, 180
  as epiphenomenon 215
  and perception 88, 90
consent, informed 132
consequentialism 224
conspiracy theory 114, 117
consumers 146
consumption of technology, worship of 143, 144–5
continuity 249
  belief in 248
  personal 255–6
control, issues of 146
convergent thinking 30–1
Cooney, Jess 53–4
cooperation 55–8, 76, 98, 117
Copernican revolution 158
Copernicus, Nicolaus 157–8
copper 15
coral cities 58
cores 53
corvids 9
Cosmides, Leda 51
cosmology 22
counterfeit people 180–1
counting 62
coupled systems 70, 72
courtship, 'sexy hand-axe theory' of 31
Covid-19 pandemic 109, 113–14, 134–5
Crawford, Kate 226
creation mythology 244
crops 14, 16
crows, tool use 9, 10, 29
cryonics 241–2
Culkin, Father Julian 3
cultural environment 125
culture 49–50, 54
  evolution 99, 110
  and optical illusions 82
  proto-human 37

INDEX

richness of 16
technological 30, 35
and technological change 18
transition 65
curiosity 5, 32, 39, 199, 205

dam-building 142
Daoism 228
Dartmouth Summer Research Project, The 203
Darwin, Charles 49
Darwinian revolution 158
Darwinian theory 99, 102
data
big 221
delusion of divine 215–38, 254
Davis, Ernest 204–6
Dawkins, Richard 113
de Beauvoir, Simone 141, 144
death 139–40, 241–2, 247–8, 251, 255–7
information-theoretic 242–3
deception 82, 90, 98, 110, 164, 168, 171
decision-making 85, 126, 128, 159, 185, 208, 254
decoding 83
Deep Blue 174, 175, 177–8, 190
deep learning 178, 204–5
DeepMind 174, 204–5
delusions 116
anthropomorphic 161–91, 253
of brutality 43–58, 252
of comprehension 97–118
delivery from 95
of divine data 215–38, 254
and illusion 88–90, 92–3, 98
of inevitability 9–23, 252
of 'it' and 'us' 59–79, 252–3
of literal-mindedness 81–95, 103, 116, 253
of machine perfection 193–213, 254
and magic tricks 164
of magical thinking 139–60, 253
manipulative/authoritarian 110

of mastery 25–42, 252
of neutrality 121–38, 253
of perpetual progress 239–54
democracy 68–9
denial 147, 164–7, 211, 254
of reality 110, 116, 117, 152
Denisovans 33, 34, 51
Dennett, Daniel 98–103, 113, 116, 180–1
Denton, Emily 232
deontological ethics 222–4
deoxyribonucleic acid (DNA) 27, 34
dependency
mutual 160, 213, 231–5
networks of 14
see also interdependence
depth 95
design, human-centred 199
design process 236
design space 99, 109
communal nature 102
desire
as evolutionary pressure for other species 14
and technology 127–8
destiny, divine 244
determinism, technological 3, 22, 52, 101, 121–2, 128–31, 231
Deutsch, David 89–90
dialogue 65–70, 156, 180, 247
diasporas, hominin 28
dictionaries 61
Didion, Joan 139, 160
'digital proctoring' software 134–5
digital technology 60–2, 77–8
Diplomacy (game) 174
disease 39, 239, 245
disinformation 111–18, 173, 184
dissent 235–8
Dobbins, Michael 144
Doctorow, Cory 200–1
dolphins 29
domestication 14

INDEX

Don, River 17
doubt 116, 207–13
'doughnut' graph 252
Dunne, John Gregory 139
dystopias 134, 150

ecology 223
economics 251–2
ecosystems 36
Edison Company 165
education 132–8, 142, 245
efficiency 211–12, 250
egoism 249
Einstein, Albert 157
electrification 165
elephants 29
Elish, Madeleine Clare 195
elites, technocratic 4–5, 145, 254
ELIZA (chatbot) 170–2
Ellenberg, Jordan 104–5
Ellul, Jacques 142–3, 144
email 127
embodiment 75–6, 180, 234
empathy 46, 55, 94, 199, 205
enchantment 163–7
encoding 83, 231
energy 36, 38, 39–40
   generation 153–5
environment
   affordances of 124, 126
   and coercion 128
   and cognitive processes 59
   cultural 125
   and the mind 92–3, 159
   natural 124–5
   and technology 98–103, 128–9
epiphenomena 215, 216
epistemic credit 71–2
equality 230, 245, 251
   see also inequality
equilibrium 246, 252
ethics 221–38
   deontological 222–4

ethical responsibility 195
meta-ethics 222
secular 229
utilitarian 222, 224–7
virtue 222, 227–9, 231, 233–5, 236–8
Ethiopia 30–1
eudaimonia 227–8
Eurasia 16, 32, 39
Europe 33, 39, 104, 162
   southern 34
Evernote 130
evil, problem of 244
evolutionary niches 27
evolutionary psychology 47, 51, 57
evolutionary theory 12, 14–15, 23, 106, 249
   and competence without comprehension 101
   and the illusion of purpose 106
   as speeding up 20, 38
   see also human evolution; technological evolution
exploration 60–2
exponential age 18–23
exponential increase 14, 15, 18, 21, 40, 97, 118, 175–6, 220–1, 253
extended mind hypothesis 59–79, 95, 159
external world
   access biases 253
   and the extended mind hypothesis 59–79
   see also reality
extinction 33, 34, 36–9, 48
   mass 41, 250
extra-neural resources 75–6
eye 85, 89
eye movement tracking 134, 135

facial recognition systems 134–7, 229–31, 237
fairness 56–7, 223, 228, 232, 234, 236, 251
   see also unfair outcomes

INDEX

faith 18, 148, 237, 243, 245–6, 249
   and science 219
   in technology as new religion 4–5, 141–5
fake news 117
fantasies 19, 22, 116, 140, 147, 152, 249
Feathers, Todd 134
feedback loops 84, 86, 91, 99, 114, 146
Ferren, Bran 1
Fertile Crescent 14
fire 10, 35–41, 43
'fire hawks' 37
Fitbits 132–3, 137–8
fitness 113
flax 14
flint 37, 53
floods, seasonal 30–1
Floridi, Luciano 158–9, 223
Foot, Philippa 194
foraging 17
forests 36
fossils 27, 33, 37
France 53–4
Frank, Thomas 211–12
Frankfurt, Harry 111
Franklin, Benjamin 162
free choice 130–1, 134
free will 133
freedom 146, 152, 159
   online 78–9
Freud, Sigmund 157–8, 159, 160
Frischmann, Brett 132–3, 137–8, 234
Frith, Chris 85
Frontier (supercomputer) 174
fuel 36, 37, 40, 41
future 152–4, 243, 250–2
   denial 254
   faith in 246
   possible 19
   sustainable 251
   unknown 249
futurology 242

Galatea 148–50, 171
galaxies 89–90
Galileo Galilei 156–7
games 105–6, 173–6, 177, 205, 208
   imitation 168–9
   see also chess
gas, natural 36
Gaughan, John 163–4
Gebru, Timnit 186
Geiger, Hans 153
Generative AIs 179
   see also ChatGPT
Generative Pretrained Transformers, ChatGPT 177–9, 182, 191
genes 12, 34, 38
   humanity's reshaping of 15
genetic variation 12
genetics 12, 33–4
Germany 11
gibbons 27
Gibson, James J. 124–5
Gigaflops 174
girls 54
glacial cycles 40–1
Gleick, James 67
Go (game) 174, 177
Göbekli Tepe complex 16–17
God 32, 244
gods 17
Goldhaber, Michael 94
'good things', replacement by 'better' 145
Google 105, 180
   augmented reality Glasses 130–1
   self-driving cars 200
Google Assistant 172
Google Translate 187
Goostman, Eugene 169
Gopnik, Alison 44, 48–50, 54, 73
Gorgias 68–9
gorillas 27
Goryeo dynasty 11
grain 15

315

INDEX

gravity 10, 22
Gray, John 244–6
Greek mythology 148–9
Greer, Evan 136–7
grief 139–40
Guggenheim museum 246–7
guidebooks 61, 62
Gümpel, Charles Godfrey 166–7
guns 122–4
Gutenberg, Johannes 10–13, 107

Haidt, Jonathan 117–18
Hall, Edith 227–8
hallucinations
    consciousness as controlled 83–5, 91–2, 159, 253
    of Large Language Models 179
hammerstones 53
hand-axes 19, 29–31
    'sexy hand-axe theory' 31
Hanna, Alex 232
Harford, Tim 107
harm reduction 224
Harper's (magazine) 211
Hawking, Stephen 216–17
Held, Virginia 235
Hephaestus 149
herding 17
Herzberg, Elaine 193–6
hidden histories 103, 105, 108, 118
hieroglyphs 64–5
Hindu tradition 93
Hoffman, Donald 84
holistic forecasting 251
Holocene (Anthropocene) 40–1
hominins 26–8, 30–5, 37, 40, 43, 51, 97, 101
    diaspora 28
    DNA 34
    migration 32, 33
Homo
    H. bodoensis 33
    H. erectus 28, 29, 32–3, 37–9
    H. floresiensis 34
    H. habilis 28
    H. heidelbergensis 32–3
    H. neanderthalensis 26, 33–4, 39, 51
    H. sapiens 4, 25–6, 33–5, 38–41, 43–6, 50–1, 97–8
hope 249, 254, 256–7
Hrdy, Sarah Blaffer 46, 47
human capabilities
    exceeded by technology 167, 176, 254
    extension through technology 146, 159
human evolution 3, 22, 25–40, 43–58, 97–102, 110, 246, 250–1
    co-evolution with technology 97, 254
    and the festival of bullshit 113
    and literacy 64
    and perception 84–5
human impact 40
human labour, hidden 190–1
human rights 78, 79
human-centred design 199
humanity
    complexity 203
    decentring of 157–60
    desires of 14, 127–8
    effort of 188–9
    uniqueness of 9–10, 23
    see also Homo, H. sapiens
Hume, David 90–1
humility 5, 118, 159, 251, 253
hunter-gatherers 14–17, 51
hunters 32–3
hyper-sociality 46, 55

IBM 174, 177
ice 41
ice ages 14, 33, 34, 36, 48, 50
'idealized machines' 198
identity 77, 79
ideograms 11

INDEX

Illich, Ivan 144–5
illiteracy 64, 65
illusions
    and delusions 88–90, 92–3, 98
    optical 81–2, 87
imagination 31–2, 58, 73–4, 97–8, 102, 152–4, 156–7, 205, 257
    narrative 170
imitation games 168–9
immortality 255–7
imperfection 254
inbreeding 39
individualism, cost-free 146
Indonesia 32, 34
Indus Valley 15
inequality 75, 114, 229–31, 233
inevitability, delusions of 9–23, 252
inevitablism 115
influencers, AI-generated 173
information, preservation 242
information age 154, 202
information environments 79, 112–15, 233
    see also informational environments
information systems 110, 113–16
information technology 67, 70, 173
informational environments (infospheres) 158–9
informational organisms (inforgs) 158–9
informational states 242
injustice 148, 210, 226, 231, 237–8, 245
ink, oil-based 12, 13
innovation 15, 23, 49, 100–1, 252
    of children 54
    and crisis 108
    deterministic accounts of 129
    and history 18
    and opportunity 108
    and perfection 254
    and social complexity 16
    worship of 144, 253
    see also invention
insufficiency, sensitivity to 56

Intel 20
intelligence 182
    animal 9, 29
    artificial artificial 190–1
    evolution 27, 32, 33
    export from humans to machines 203–4
    machine superintelligence 4, 22, 183–4, 193–213
    tests of 168
    see also Artificial Intelligence
intention 112–13, 115
    attribution to artificial agents 190
interconnectedness 110, 117, 129, 146, 158
interdependence 3, 48, 228, 236
intergenerational care 58
intergenerational dynamics 49–50, 239, 243, 247, 250–2
intergenerational justice 251
interiority 92, 215–16
    see also mental life
Internet 78, 177
introspection 92, 215–16
    see also mental life
invention 2, 10, 61, 66, 108, 128, 131, 240
    see also innovation; self-invention
iPhone 174
'it' and 'us', delusion of 59–79, 252–3
Italy 13
iteration 10, 99–100

jays, pinyon 55
Jesus 67, 229
Johansson, Sverker 34
Joseph II, Archduke 162
justice 149–50, 185, 223, 228, 247
    intergenerational 251

Kahneman, Daniel 76
Kant, Immanuel 222
Kasparov, Gary 174–8, 190
Kelly, Kevin 125–6, 142

INDEX

Kempelen, Wolfgang von 161–4, 166–8, 190–1
knowledge 118, 136, 205–6
   new 157–8, 233
   see also self-knowledge
Kranzberg, Melvin 122
Krznaric, Roman 251
Kuhn, Thomas 155–7
Kurzweil, Ray 21–2, 175–6, 177, 220

LA Times (newspaper) 183–4
labour, hidden human 190–1
lactose 51
Lancet, The (journal) 95
language 58, 69–70, 171, 181, 189–90
   acquisition 63
   as data 187
   Sumerian 64
Lanier, Jaron 114–15
larch 17
Large Language Models (LLMs) 177–80, 182, 184, 186, 205
Latour, Bruno 123
Le Guin, Ursula 116–17, 123, 147–8
lead 11
learning 49, 76, 205–6
   from accidental revelation 108–10, 201
   machine 182, 188, 211, 221, 230
   remote 134–5
left occipito-temporal cortex 64
Lemoine, Blake 180
lemurs 46
Levant 15, 16
liberty 228
Libin, Phil 130–1
Libratus 174
lies 111, 116
life 241–2
life-enhancing technology 239–40, 245, 254
limestone T-shape pillar complexes 16–17

literacy 63–5
literal-mindedness, delusion of 81–95, 103, 116, 253
LLMs *see* Large Language Models
Loebner Prize 169
Logic(s) (magazine) 232
longtermists 226
love 43, 48, 49, 52, 235–8, 256–7

M2 Max (chip) 20
MacAskill, William 226
McGilchrist, Ian 92–3, 95
machine learning 182, 188, 211, 221, 230
machine perfection, delusion of 193–213, 254
machine superintelligence 4, 22, 183–4, 193–213
machine-mediated world 213
machine-optimized environments 198–9, 212
machines
   attribution of autonomy to 164
   brain as 86
   and control issues 146
   'idealized' 198
   mind and 253
   objectives 207–11, 213
   rise of the 146
   and salvation 131, 241, 246, 254
   undying 251, 254
MacIntyre, Alasdair 118, 233–4
McKenna, Sioux 135–6
McLuhan, Marshall 3, 129
McMillan-Major, Angeline 186
magic 163–6, 169
magical thinking 173
   delusion of 139–60, 253
Maimonides 228
maps 60–1, 62, 71
'March of Progress, The' 25
Marcus, Gary 204–6
Maria Theresa, Archduchess 161–2

INDEX

Marsden, Ernest 153
Marvin, Carolyn 165
mastery 48, 117
   delusions of 25–42, 252
materialism, promissory 219, 220
mathematics, applied 221
Mechanical Turk 161–4, 166–7, 190–1
Mediterranean 67
meerkats 55
Meliorism 244
memes 113–14
memory 62, 66, 242
Memphis 15
menopause 47
mental life
   abolition 215–16
   and the construction of technologies 30
   description and replication 92, 203
   emergence of 97
   mapping to the physical realm 30
   see also interiority
mental models 84, 93
mental processes 10, 43
mental time travel 19, 31
Mercedes-Benz, Drive Pilot system 200
Merchant, Brian 183–4
Mesopotamia 17, 64
Meta 174
meta-ethics 222
metal alloys 11–12
metal smelting 10–11, 13, 15, 40
metaphysics 243
Metaverse 130–1
microchips 20
Middle East 14
Midgley, Mary 219
migration 32, 33
mind 51–2, 55, 57, 81, 110, 146, 246
   boundaries of the 59
   and consciousness 180
   decentring of 158

emergence 98–9
and the environment 92–3, 159
evolution 31
extended mind hypothesis 59–79, 95, 159
hive 57–8
and machines 253
manipulation 115
'mini-me' view of the 84–7, 91–2
Palaeolithic 54
plasticity 64, 97
as story processor 117
technological enhancement 146, 159, 181, 254
technology as aspect of 63, 70–5
theory of 62
as theory-laden 98
minority social groups 49
mobile phones 60–1, 71, 72, 74, 78
modernity 50, 51, 109, 129, 236
   technological 74, 165
monitoring, passive 132–8
monkeys 27, 55
   capuchin 29
   langur 46
   red colobus 46
Moore, Gordon 20
Moore's law 20–1
moral character 228–9
moral choice 128, 131
moral duty 222–4
moral labour 128–9, 133, 234
morality 225, 235
Morozov, Evgeny 143–4, 145
Motherboard website 134, 135
motion 155, 156
Müller-Lyer, Franz Carl 82
Müller-Lyer illusion 87
multiplicity 48, 206, 208, 233
Mumford, Stephen 217–18
Muslims 13
mutation 12
mutual care 46–7, 54–5, 58, 233

319

mystery 143, 144, 148–52
myths 142, 147–52, 159, 219
   creation 244

Napoleon Bonaparte 162–3
narrative imagination 170
narratives (stories) 15–16, 18, 106, 116–18, 139, 145–52, 157, 205
   about technology 145–8
   anthropocentric 102–3
   and data 219–20
   of inevitability 237
   of optimization 237
   see also story-telling
National Transportation Safety Board (NTSB) 193–6
natural selection 38
Nature Machine Intelligence (journal) 223
Neanderthals (*Homo neanderthalensis*) 26, 33, 34, 39, 51
needs 13, 22, 57, 127–8
Neolithic people, Early 17
networks 146
neural networks 178, 205
neuroplasticity 44, 64, 76, 97
neuroscience 87
neutrality, delusion of 121–38, 253
New York Times (newspaper) 94
New Yorker, The (magazine) 189
Newton, Isaac 156–7
Nicole, Hilary 232
non-maleficence 223
normality, and injustice 237
normalization
   of algorithmic data processing 137
   of surveillance 137, 237
norms, of science 217
Nowell, April 52–3
NTSB *see* National Transportation Safety Board
nuclear energy 153
nuclear physics 153, 154–5
nuclear weapons 153, 183
nurture 44, 49, 229, 234–5, 250
Nye, David 198

objectives, human versus machine 207–11, 213
observation 216, 217–18
obsidian 30–1, 53
obsolescence, human 176
Oceania 34
octopus, Indonesian 29
online freedom 78–9
opportunity 79
optical illusions 81–2, 87
Oral Roberts University 132–4, 137–8
oral tradition 65–70, 239
orangutans 27
oratory 65–70, 239, 247
orcas 9
Ord, Toby 226
Organisation for Economic Co-operation and Development (OECD) 65
otters 29
Ovid 148–9, 151, 171
oxygen 36, 41

Palaeolithic people, differences of 50–4
Palomar Sky Survey 89–90
paper 12, 13
'paperclip maximiser' 209
Paranthropus 28
parenthood 234–5, 255–6
Parfit, Derek 226
Paris 162
participation
   and Artificial Intelligence 185, 200
   and technological progress 145
   in written and recorded culture 65, 67–8
particle accelerators 154
passivity 132–4, 137–8
pasts, parallel 19

INDEX

Paul, Grand Duke of Russia 162
Paul, Annie Murphy 77–8, 79
peas 14
perception 83–93
    and 'bottom up' processing 83–4, 86
    and inner control models 84, 85
    integration with technology 88
    meaningful test of 87–90
    and reality 81–2, 86–93, 98
    as theory-laden 98
    and 'top down' processing 84–5, 86
perfection *see* machine perfection
perpetual progress, delusion of 239–54
personal experience 248–9
personal growth 237–8
personality 242
Phaedrus 65–6, 68
philosophical dialogue 65–70
philosophy 215–18, 222, 257
Phoenicians 64–5
photosynthesis 41
phronesis (practical wisdom) 237
pigs 14
Plato 110, 239, 247
    Gorgias 68–9
    Phaedrus 65–8
    The Republic 69–70
Pleistocene 71, 74
Popper, Karl 219, 220
population growth 22–3
post-reproductive adults 47
posthumanism 243–4
power 145, 254
praxis 237
prediction 153–5, 159, 167, 190, 198–9
prefrontal cortex 44
prehistory 14–18, 50–2
prejudice 231
primates 9, 27, 29, 45–7, 56
    see also specific primates
primitiveness 15–16
printing press 10–13

privacy 223
process theologies 244
profit 13, 94, 113, 123, 129, 190, 199, 201, 254
progress
    actual 241–6
    damage done by 108
    delusion of perpetual 239–54
    imagined 241–6
    and solution-seeking 101–2
proto-farming 17
proto-humans 37, 51
psychology
    evolutionary 47, 51, 57
    as objective science 215–16
psychotherapy, automated 170
purpose 221
    divine 244
Pygmalion 148–9, 151
Pyne, Stephen J. 40–1
Pyrocene 40

race 229–31
Raihani, Nichola, *The Social Instinct* 55
Raritan (magazine) 111
rationalism 219
rationality 117, 209
rationalization 102–3
Raworth, Kate 251–2
reading 63–4
reality 245, 245–6, 254
    and Artificial Intelligence 177–81, 188
    as blinding to future possibility 153
    complexity of 188
    denial of 110, 116, 117, 152
    flux of 93
    human access to 81–2
    ignoring 115
    loosening the collective grip on 181
    narrative reflection of 118
    and perception 81–2, 86–93, 98
    sense-making of 98

INDEX

reality – *Cont.*
   and technological dreams 194–9
   uncertainty of 188
   viciousness of 106
   see also external world
reason-giving 102–3
recording 65–8
   active 132–4, 137–8
records 65, 67–8, 70
regionalism 142
religion 244
   innovation as 253
   technology as 4–5, 141–5
remote learning 134–5
reproduction, sexual 12–13
resilience 44, 115, 250, 254
resistance 237–8
responsibility 152, 186, 223, 236
   ethical 195
   legal 195
   transfer from the individual 134, 145
   wishful abnegation 4–5, 123–4, 130, 145
retina 85, 89
revelation, accidental 108–10, 118, 201
rewards, relative/absolute 55–6
rhetoric 63–70
Rid, Thomas 146
right action 222, 224
Riley, Henry 149
risk reduction 224
road traffic accidents 193–5
Rock of Gibraltar 39
role models 228–9, 237
Roman alphabet 12
Rorty, Richard 58
Rose, Janus 134
Rouffignac cave system 53–4
Rovelli, Carlo 156
Russell, Stuart 175, 207–8, 210
Russia 17
Russian roulette 105–6
Rutherford, Ernest 153, 154–5

Sacasas, L. M. 129–30
Saint-Acheul 29
salvation
   and efficiency 250
   machine-made 131, 241, 246, 254
Scheffler, Samuel 247–9
Schmidt, Klaus 16–17
Schwarzenegger, Arnold 150, 151
science 216–20, 245
   as act of faith 219
   norms of 217
scientific revolution 154–5, 158–9
sea levels 40
seances 165
Second World War 104–5, 256
secularism 244, 246
Seedol, Lee 174
Sehgal, Tino, *This Progress* (2006) 246–7
self
   'bundle' theory 91
   dynamic, divided 90–3
   in state of flux 93
   work on the 229
self-assessment 132–4
self-enhancement 239–40
self-invention 2, 23
self-knowledge 54, 55, 95, 206, 236
   uncertainty of 158, 159–60
self-perception, paradox of 91
selflessness 249
Selinger, Evan 132–3, 136–8, 234
Semitic languages 64–5
sense-making 76–7
sensory inputs 83–7
September 11th attacks, 2001 111, 256
Seth, Anil 83, 85–8, 91
sharing 55–6
Shaw, George Bernard, *Pygmalion* 171–2
sheep 14
Shmitchell, Shmargaret (Margaret Mitchell) 186
Shogi 174

INDEX

Sicily 68
side effects 109
simple-mindedness 225
simplicity 93
simulation hypothesis 243
Singer, Peter 224–6
Singularity 21–2, 152, 176–7, 220, 243–4
Siri 172
skull 45
Smart, Andrew 232
Smart, J. J. C. 225
'smart' technologies 200
smartphones 174–5
sociability 46–8, 55–6, 75
social change 230
social comparison-making 55–6
social complexity 16, 33
social engineering 169–70, 173
social 'fads' 9
social robots 229–31
Society of Automotive Engineers, The 199–200
Socrates 65–70
Solomon, Andrew 255–6
solution-seeking 101–2
song 100–1
soul 92
Spain 13
Spenser, Edmund, *The Faerie Queene* 149–50, 151
Standage, Tom 163–6
Star Trek: The Next Generation, Borg Complex 129–30
StarCraft II 174
stars 89–90
Stephanopoulous, George 111–12
Sterman, John D. 109–10
Stilgoe, Jack 197–8, 199
stone tools 9, 38, 43, 101
  hand-axes 19, 29–31
  and novice tool-makers 52–3
story-telling 183–4

consciousness as form of 215
  and data 219–20
  see also narratives (stories)
subjectivity 216, 247
subscription services 200–2
Suddendorf, Thomas 31
Sumer 15
Sumerian language 64
sun 36
supercomputers 152, 174
superforecasters 154
superhuman competence 204–5
superintelligence 193–213
  see also Singularity
supervolcanoes 34
surveillance 132–8, 200–2, 222–3, 235–6
  normalization 137, 237
survival 12–13, 16, 28, 34–8, 47, 49, 55, 98, 100, 105–6, 110, 113, 115, 248–9, 253
survival of the fittest 49, 249
survivorship bias 103–8
symbiosis 12

Taleb, Nassim Nicholas 105–6
Talos 149–50
Tasioulas, John 185
Taylor, Charles 173
Taylor, Timothy 45
techno-social engineering 132
technocracy 4–5, 137
technocratic elite 4–5, 145, 254
technological culture 30, 35
technological determinism 3, 22, 52, 101, 121–2, 128–31, 231
technological evolution 110
  competence without comprehension 101
technological modernity 74, 165
technological progress, as demanding expert control at the expense of participation 145

INDEX

'technological sublime' 198
technology 1–5
  anxiety-provoking nature 67
  apotheosis 177
  as aspects of our minds 63, 70–5
  assumptions embedded in 122–3, 125, 127, 132–8
  children and 77
  co-evolution with humanity 97, 254
  and coercion 128
  and common inheritance 117
  and competition 117
  complexity 14–15, 18, 20, 101, 202, 220–1, 224, 225, 253
  and conquest 117
  consumption 143, 144–5
  contradictions 207
  and cooperation 117
  and culture 117
  definition 1
  and the delusion of anthropomorphism 161–91, 253
  and the delusion of brutality 43–58, 252
  and the delusion of comprehension 97–118
  and the delusion of divine data 215–38, 254
  and the delusion of inevitability 9–23, 252
  and the delusion of 'it' and 'us' 59–79, 252–3
  and the delusion of literal-mindedness 81–95, 103, 116, 253
  and the delusion of machine perfection 193–213, 254
  and the delusion of magical thinking 139–60, 253
  and the delusion of mastery 25–42, 252
  and the delusion of neutrality 121–38, 253
  and the delusion of perpetual progress 239–54
  and the denial of reality 110
  and divinity 21
  emergence 29, 31
  and the environment 98–103, 128–9
  evolution 3, 10–12, 15, 20–2, 38, 101
  expecting too much from 199–202
  and exponential increase 14, 15, 18, 21, 118, 175–6, 220–1, 253
  faith in 141–5
  and *Homo erectus* 28
  inability to save us 22
  informed negotiation with 82–3
  integration with perception 88
  iterative nature 99–100
  life-enhancing 239–40, 245, 254
  and mastery 117
  mind-enhancing 146, 159, 181, 254
  momentum of 126, 127
  and narratives 117
  needs of 13, 22
  as new religion 141–5
  and new tech as a recombination of older tech 10–12
  origins 7–118
  seductive 167–73
  self-design/improvements 21
  'smart' technologies 200
  in stasis 19
  and survival of the fittest 49
  transformation of tools in to 28–32, 97
  transmission 12–13
  unintended side effects 109
  virtuous cycles of 227–31
  wants of 125–7, 142
  see also Singularity
technosocial opacity (blindness) 221, 223, 227, 236
Tennessee Valley Authority (TVA) 142, 143

INDEX

Terminator 2: Judgement Day (1991) 150–2, 172–3
termite mounds 58
Tesla 166
  'autopilot' system 200
Tesla, Nikola 165–6
Tetlock, Philip 154
Texas Hold 'Em (game) 174
theology, process 244
think-and-record activities 132–4
This Week (magazine) 111–12
Thomist Christianity 228
Thompson, E. P. 18
thought
  articulation 189
  and the external environment 79
  and language 190
  see also cathedral thinking; convergent thinking; magical thinking
thought experiments 194, 209, 247–8
thriving 2, 27–8, 33, 35–6, 41, 43, 55, 58, 73, 78, 100, 110, 234, 249, 251
time 20, 105–6, 154–5, 249–50
Time (magazine) 191
time travel, mental 19, 31
tin 11, 15
tipping points 21, 42, 243
Toba, Lake 34
tombstone mentality 197–8
Tooby, John 51
tools 3, 4, 9–10, 71–2
  Acheulean toolmaking 29–31
  animal use of 9, 10, 29
  biological 14
  and control issues 146
  crude 9
  'hafting' technique 38
  hand-axes 19, 29–31
  hominin use of 28, 43
  and the necessity of their maker/maintainer 18–19
  neutral tool concept 122–8, 130
  and recursive iteration 10
  stone 9, 19, 29–31, 38, 43, 52–3, 101
  transformation into technologies 28–32, 97
  see also stone tools
'top down' processing 84–5, 86, 205–6
totalitarian regimes 183
totems 17
trade routes 15
training 182
training sets 230
transcendence 54, 57, 242, 246, 251, 254
Transformers, ChatGPT 177–9, 182, 191
transhumanism 242–3, 246
transistors 20–1
translation services 187
transparency 223
transport 122
travel 60–2
trickery 161–73, 180–1
Tripitaka 11
trolley problems 194
'trouble-free' worlds 145
Trump, Donald 111–12
truth 111, 115–17, 156, 180, 254
  indifference to 110
truth-seeking 115
Turing, Alan 167–9
Turing test 167–73
Turk 161–4, 166–7, 190–1
Turkey 16–17
Turkle, Sherry 173, 202
TVA *see* Tennessee Valley Authority
Twitter 230

Uber 193, 194–7
uncertainty 116, 152, 188, 223, 233, 249
unconscious 158
  collective 62
unfair outcomes, aversion to 56–7
United Nations Educational, Scientific and Cultural Organization (UNESCO) 65

325

INDEX

United Nations General Assembly 78
United States 42
universe 22
  humanity's place within the 157–60
Upanishads 93
Upper Nile 15
Urals 17
uranium fission 153
utilitarian ethics 222, 224–7
utilitarianism 224–7, 232

Vallor, Shannon 128–9, 133, 221, 222–3
values 122–4, 133, 136, 233, 253
Van Gelder, Leslie 53–4
Vasquez, Rafaela 193–5
vehicles, autonomous 193–8, 199–202, 212–13
Véliz, Carissa 237–8
Venus 148
Versailles 162
Vice (magazine) 134
  see also Motherboard website
video playback 88
Vienna 162
Vince, Gaia, *Transcendence* 36, 38, 39
violence, human capacity for 47
Virilio, Paul 108–9, 117–18, 127, 201
virtual assistants 172–3, 187
virtue ethics 222, 227–9, 231, 233–5, 236–8
virtuous processes 227–31, 254
viruses 113, 115
vision, persistence of 88

visual word-form system 64
Volkswagen 200–1
Volvo SUVs 193, 195–6
vulnerability 2, 4, 33, 38, 43, 48, 54–5, 58, 97, 114, 173, 233–5, 252

Wald, Abraham 104, 106–8
wants, of technology 125–7, 142
Warzel, Charlie 94
Watson, John B. 215–17
Waymo 200
Weinberger, David 187–8
Weizenbaum, Joseph 170–1
well-being 128
wheat 14
Williams, Bernard 225
wine-presses 11–12, 13
Wired (magazine) 218
wish-fulfilment 139–41, 145–6, 153, 172, 237
women, absence from the technology industry 49
wood block printing 11
woolly mammoths 17
World Trade Center 111
world-states 31–2
worldviews 110, 144, 157, 159, 184, 244, 248
writing 62, 63–70, 181, 239

Zallinger, Rudolph Franz, 'The Road to Homo Sapiens' 25
zero sum propositions 94
Zoom 230